A TAPESTRY OF LIFE
AN AUTOBIOGRAPHY

JAMES S. CANDY

MERLIN BOOKS LTD.
Braunton Devon

First published in Great Britain, 1984

ISBN 0 86303 188-9
Printed in England by Short Run Press Ltd., Exeter, Devon

FOREWORD

I first met James Candy at Youlbury on Boars Hill above Oxford in the summer of 1984. It was one of those Oxford summers when, viewed from Jarn Mound, the city seems to lie before you in its bowl of hills shimmering in a liquid haze. On such a summer seventy years before, James had received the news of the beginning of the Great War in the company of Sir Arthur Evans, the excavator of Knossos. I had come in search of the real Evans for a BBC documentary series, for James had actually lived in Evans's house here during those last years of the Edwardian era; though a boy, he knew the great man in a way his fellow scholars never could. And Evans *was* a truly great man, not only one of the most remarkable scholars this country has produced but also adventurer, foreign correspondent, archaeologist, prehistorian, and expert in so many related fields, such as art history and numismatics: a wealthy man who devoted himself (and his fortune) to scholarship. It is Evans whom today's tourists celebrate when they walk in their thousands around the great Cretan site of Knossos: as they walk its courtyards, descend its Grand Staircase, enter its throne room, they are in a sense entering into Evans's imagination, for these are his reconstructions. It is too late now to recapture much of Evans's many sided genius, but James Candy is our last living link with the man himself, and that is the first reason why James Candy's charming biography is so welcome.

Evans built Youlbury for his wife Margaret, daughter of the great Oxford historian E. A. Freeman. She died tragically young in 1893, but he went ahead with his plan – the landscaped gardens, the artificial lakes, the house itself, its viewing platforms across the Vale of the White Horse to the Berkshire Downs. Youlbury remained her monument and there Evans lived, when not in Knossos, till his death in 1941. He never forgot her: to the end his notepaper was black-edged with an elephant's head as James tells us. The house is gone now – replaced by a modern pile – but the gardens Evans planted remain, now of course far more lush and mysterious than when he saw them, a truly secret garden with tangled paths overhung with pink and white rhododendrons under a canopy of oak and pine. Rarer plants – Himalayan poppies or the 'strawberry' tree – bespeak Evans's intense love not only of flowers but of all the natural world: 'he could tell you as easily of the life on the bottom of the lake as of the archaeology of Knossos' says

James. Walking through those woods at James's side as he conjured up those lost conversations from those far away summers, Evans came vividly to life as a man: there he was, sometimes casual with an old white jacket, open-necked shirt and a huge heliotrope drooping from his buttonhole; sometimes dapper with trimmed moustache, suit and natty trilby, taking the boys out on a picnic to Snowdon or out to Wayland's Smithy by car (though not a driver, Evans notoriously loved fast cars); or best of all (as James relates in his book) playing the Minotaur at his Twelfth Night parties for all the local boys and girls – bellowing from the centre of the black and white marble labyrinth in the floor of the hall at Youlbury, to the delight (and occasional terror!) of the younger ones.

Evans never had children of his own; that is why he became James's guardian. In its turn this is what makes James's account of their relationship so touching. For though he had the reputation of being stern to his fellow academics, Evans evidently had a marvellous ability to enthuse and encourage children, to speak their language, and thrill them with love of the natural world and love of the past; he also was tremendously kind, which comes out in a number of little incidents related by James. For the journalist and film maker it is frustrating now, forty years or so after Evans's death, to find that our radio archives have no preserved interviews with the great man, only one much later TV interview with his late half-sister Joan, who wrote his biography. It is our loss: from the picture which emerges in James's book one imagines Evans would have been a must for TV, rather like Sir Mortimer Wheeler. This loss gives James Candy's memories their special interest.

For the BBC team who worked with 'Sir James' (as we called him!) our days at Youlbury were a delight; we recorded our interview on the bank of the lower lake, by Evans's old bathing huts and a waterlogged punt – a magical place, frequented now only by a pair of swans. As James brought back Evans's spirit with increasingly uncanny detail, we turned at one point, almost expecting to see that wiry figure coming bustling through the gate, scratching the back of his head (as he was wont), waving Podger (his favourite walking stick) and shouting 'er-er-Jimmy! Who are these people and what are they doing?' As we walked back through the wood Evans's lonely and restless spirit seemed even closer, lost in this English version of the idyllic world he had created in the garden of the Villa Ariadne at Knossos, that one inhabited by more ancient ghosts. Before we parted James recalled the last days in the spring of 1941:

'Towards the end I was married and up to my eyes in work. But there were times when I'd come up here on my own, and I'd look through the window into the big drawing-room. There he was sitting in the chair with the rug over him and the black cat sitting there beside him. Entirely on his own. He said to me one evening, "I ought not to be living here Jimmy." I said, "Why not Sir Arthur?" He said, "You know, I've spent all my money on Knossos." And so he had.'

But if Evans's story gives James Candy's book its particular interest, let's not forget James's own story. His intimate friendship with Evans was as a boy – though they corresponded regularly through the years which followed. But when James grew up he left these shores for South America, to make his fortune. There his adventures in the pampas of Argentina are at times real 'Boys Own' stuff, with James meeting all problems – from knife-throwing gauchos to houses of ill repute – with the same delightful humour and equanimity! The story of how he met his wife-to-be is a little romantic classic of its own.

Now James lives in well-deserved retirement in Abingdon, his garden, perhaps (or so it seemed to me) an echo of those magical glades at Youlbury in its beauty and variety. Happy and tranquil, and irrepressibly humorous, 'Sir James' in this book reminds us of those older times not with nostalgia but with a natural feeling for what is good in life. And these days that is no bad thing.

Michael Wood

ACKNOWLEDGEMENTS

I should like to express my gratitude to Mrs D. Thursfield for her wonderful help during the writing of this book, and my appreciation to my daughter Jenny Benson, and Mrs Sylvia Horwitz for their help and encouragement.

J.S.C.

I have had many requests from my family and friends for me to write some sort of autobiography of the many interesting experiences that I have had during my life. This undertaking has to have a beginning and an end. That being the case, I shall briefly cover my ancestors as far back as 1550.

A well-known family by the name of Yeoman was living at Wanstrow, Dorset, at that time. Several generations later, a certain John Yeoman married Ann Harding around 1807. From this union a daughter named Nancy was born and she married William Candy, which, in due course, brings me to my own birth in 1902.

It is only proper to look into the family name of Candy which was well known in the West Country, especially Somerset and Dorset. The earliest report of Candy with its variants, Candie and Candey, is stated to come from Cande near Blois (there is a Candé north of Angers on the railway line between there and Nantes). The name of Nicholas Candy occurs in Normandy in 1195. Then in England the name is recorded of John Candy of Suthewic who held all the rights in a water-mill in the village of Stepelmorden in August 1318. Sometimes it was spelt Candey or Cundey and in 1406 there is a reference to John Cande.

On the other hand, the name of Candy is of French or Flemish extraction and may be derived from Candè or Candé, who were refugees from religious persecutions after the massacre of St. Bartholomew in 1572. By 1585 in England, the spelling was fixed at Candy and has not changed.

We now come to the very small village of Cloford in Somerset and it is from there at Cloford Manor, we see over the porch, 'Thomas Candy, 1795'. For many generations the name of Candy has been associated with farming and it was in 1973 that the last farming Candy, whose ancestors were from Cloford, died in the village of Boars Hill in Oxfordshire. There is only one Candy descended from that long line from Cloford farming today; he is Field Candy in New Zealand.

Cloford Manor, not far from Shepton Mallet, close to Wanstrow, is a delightful old manor with mullioned windows. It remained in Candy hands until it was sold, with the farm, in 1900. At the present moment, the owner lives in a new house he built himself and has turned the old manor into a corn store.

Thomas Candy, who built the house, was asked by the parishioners if a wide passage from his back door to his front door could be made to enable coffins to pass through to reach the church which was some quarter of a mile from the manor. Otherwise, they would have to make a long detour by road of two miles to reach the church. This Thomas Candy agreed, with the proviso that the occupants of the coffins must have lived and died in Cloford village. It was considered as a right-of-way for funerals only, but now, in modern times, that has long been forgotten. The wide passage and front door which allowed the coffins to be carried through, can be seen to this day.

One amusing incident must be recalled concerning the Candys at the manor. In about 1860 there were farming at Cloford, two bachelor brothers, looked after by an elder sister. It seems one day, the sister roasted a hare for the midday meal. On sitting down at the kitchen table, the two brothers complained bitterly at having to eat hare, yet again, for lunch. Their sister took offence at their complaints and, being rather quick-tempered, picked up the hare, threw it at her brothers, rose from the table and flounced out of the room to her bedroom. The two brothers, I suppose, ate the hare in the end, for it was that or nothing. Afternoon milking came round so off they went to the cow shed. Late afternoon, after the stock had been bedded for the night and the other jobs done, they returned to the house, expecting to see their supper all ready for them. To their surprise, there were no signs of a meal or their sister and they thought she must be sulking in her bedroom.

Time went by and one of them said he would go upstairs to persuade her to come down. When he received no reply to his knocking, he opened the door and went in. On seeing the state of the room, he called his brother to come up at once. They saw the empty drawers and cupboards and noticed that a small trunk had been taken. This indeed frightened the brothers, for it was obvious she had left.

On making inquiries in the village, they heard that their sister had been taken by pony trap to the railway station at Shepton Mallet. There, the station-master revealed that she had bought a ticket to Liverpool. Now, Liverpool was a long way away and farming and milking cows could not be left, and it was only after some weeks, they found out that she had booked a passage to New York.

Not a word was heard from her, so imagine their surprise when one day, after twenty years, she turned up at the manor. This indeed caused a stir in the village, as the brothers had given her up for dead. Not a bit of it, for it seems she had got a job as a domestic in New York and then had become housekeeper to a rich lawyer and, in turn, his wife. Now she was a rich widow and thought it would be a good thing to return to Cloford and look up her two brothers. It seems she only stayed a few days at Cloford. She returned to New York and her brothers never saw her again. There were no children from that marriage. I wonder what happened to her money!

James appears in every branch of the family tree. Large families were the order of the day and if the eldest son who, it seems, was often called James, died soon after birth, then the following son was called James. In fact, I have counted over twenty Jameses in the family tree. It is quite amusing to visit Cloford church and see three James Candys. The first was in 1822, aged seventeen years, the second in 1834, aged sixty-six years, and the third in 1870, aged fifty-seven years. He was churchwarden, and there is in the church a nice stained glass window to his memory. Also buried at Cloford are :

Christopher Candy in 1715,
Thomas Candy in 1798 aged 57, and
Elizabeth Candy in 1821 aged 45.

In the Yeoman family from Wanstrow, from which I am ultimately descended, was a Mary Yeoman who married, in 1780, a Joseph Harding. Their son, Joseph, was the inventor of the process for making Cheddar cheese. Mary, before she married, kept a diary which was found in Australia and is now in my possession.

Now we are coming more up-to-date. From the marriage of William Candy and Nancy Harding, in 1837, were born twelve children and the second son, Walter Yeoman Candy, who married Sara Wigmore was my grandfather. He died in 1926. He, too, had twelve children and my father, George Candy, was the eldest. He married Ellen Hedderley in 1886 and this is where I come into the picture. I was born in Boars Hill in 1902 and was their third son.

My grandfather, always a tenant farmer, drifted from one derelict farm to another, always on the verge of bankruptcy. My father, the eldest son, was born in Corfe Mullen; another son at Wimborne, and a third at Tyneham. He was no farmer, but he certainly was very prolific. His poor wife had twelve children and not one of them died in infancy, as happened in so many families. Strange to say, only one of his twelve children did not reach the age of eighty. The others, including my father, lived well into their eighties. One uncle was ninety-four and the other ninety-six before they died.

The last farm that my grandfather rented was at Sway, near Bournemouth, where I spent the summer holidays. From all accounts, he was very fond of brandy, but my grandmother refused to have any alcohol in the house, so grandfather kept a bottle hidden in the chaff house, from which he frequently had little nips! He had a habit, after taking too much brandy, of wanting to have a row with his wife, who always managed to get the better of him when he was sober. He started by rubbing his right thumb nail up and down his waistcoat buttons and this was an indication to his children to 'get lost' as soon as possible. He was also very fond of potatoes, left over from the previous day, fried for his breakfast and, if they were not just to his liking, he would throw them out into the garden and make his poor wife start all over again. He would be considered a cruel man today.

My father and all his brothers had to get up at 5 a.m. to milk the cows before going to school. Woe betide any one of them if he fell asleep against the warm flank of a quiet cow on a cold winter's morning, for my grandfather would throw a milking stool at the cow! All my uncles, including my father, left the village school when they reached the age of twelve and they all escaped from the farm, as soon as possible. The nest did become rather overcrowded.

In due course, because of old age, they had to give up farming and they retired to Kennington, near Oxford, where my Uncle Will, having made a lot of money exporting pedigree cattle to the Argentine, kept them in comfort.

My grandfather and my grandmother were both born in 1840. Walter Candy died in 1927 and Sara, his wife, in 1916. I used to visit my grandfather after his wife had died, at Kennington, where he was looked after by Nancy Wigmore, a cousin who had never married. He used to sit in front of the fire twiddling his thumbs. One day I asked him why he did that. His reply was as follows:

"Sometimes I sits and thinks and other times I just sits." At the time, I thought this very profound. Grandfather also referred to all food, whether for himself or his cattle, as 'vittals'. "Where's me vittals?" he would shout, on entering the house and seeing nothing on the table.

You will find the name of Candy in any part of England, by looking in the telephone directory, but I like to think I am descended from the old Cloford family. My father, was born in 1863 and my mother, a Hedderley of Oxford, in 1862. They both died in 1948 at Harwell in Berkshire.

When my father, George Candy, was aged twenty, he decided to leave for 'foreign parts' and took a ship bound for Montevideo, the capital of Uruguay. On landing there, he received a cable saying that the last of the batch had arrived, this being my Uncle Frank. He was the 'dillon' of the family, never married and spent his whole life working on farms. He died at the age of eighty-four at Middle Farm, Wootton, Oxon.

My father soon moved to Buenos Aires and landed, as one had to in those days, by getting ashore in a small boat. For some time he tried to get work but was unable to do so because he could not speak Spanish. He soon made friends with a Scottish family, called Macdonald, who lived, as did other British families, in the suburbs of Buenos Aires and who had arrived to construct and run the British-owned railways.

My father, through his friend Mac, was offered a job as fireman in the goods train that he drove. Now, my father loved playing card games of any description and would drop anything to do just that. It turned out that Mac, too, loved playing cards so when they left B.A. for the southern region to pick up cattle and grain, they soon found out that at certain stations, the station-masters had the same passion for cards. In those days, the tracks were single with, perhaps, one passenger train a week, so, not having a tight schedule, the engine driver told my father to fire the train as hard as possible to enable them to reach a station of their choice, for a good game of cards for an hour with the station-master! On one occasion, on a passenger train, the engine driver's cap blew off and my father carefully noticed the spot where it landed and the following week, stopped the train and retrieved the cap.

It was soon after this that my father met a French Basque, who had bought an estancia some distance from Balcarce, a small village, where he wanted to produce milk, etc. This was just up my father's street, and soon they became partners; my father putting the 'know how', i.e. milk and butter production and general running of the cattle, etc. into the business. It was not long before my father gave Recarborda, his partner, the idea of making Cheddar cheese, as one of his ancestors had first made it in England. Father was despatched to England to purchase machinery, etc. for the factory which was to be built while he was away.

While in England and on a visit to see his cousin, Frank Wigmore, who owned a dairy business in Oxford, he met the Hedderley girls. One of them was to become my mother, Ellen Hedderley, a very pretty girl, who had just finished her education in Germany. They were soon engaged and, meanwhile, Frank Wigmore proposed to, and was accepted by, another Hedderley girl. The result was a joint wedding in Oxford. Most people thought that my father's prospects in Argentina would surpass Frank Wigmore's dairy business in Oxford. How wrong they were and how differently it all turned out.

During their honeymoon, my father and mother were out walking when my father spotted an old card-playing crony going into a pub, so he told my mother he would like to speak to his old friend and would she wait for him; he would not be long. Father was soon inveigled into playing cards and forgot my mother. Poor mother! She waited about for some time and then returned to the hotel and in due course he turned up. History does not say what she said, but I can guess. Soon after the wedding they sailed for B.A.

My mother was flabbergasted when she saw the factory and her new home – everywhere flat land as far as the eye could see, very few trees, miles of fencing and an odd windmill here and there. Her new home consisted of mud walls and floors, and a tin roof. But worst of all, there were millions of flies. The nearest village was miles away and the only means of transport was on horseback or by buggy.

Father brought with him three Irish families to work in the factory on butter and cheese making. It was not long before they started quarrelling amongst themselves, so he split them up by building their houses five miles apart from each other. That did not stop them, for they all met on Sundays just to have a good row! In the end, he sacked the three families.

When my mother was pregnant for the first time, there being no medical facilities or doctor, my father set off with her in a bullock wagon to B.A., two months before the expected birth. It took three days, with no apparent bad effects. My eldest brother, Gilbert, was born in 1889. He died in New Zealand in 1982, aged ninety-three.

My father, having established himself at Balcarce, was soon followed by ten of his brothers and sisters at different times, all anxious to be liberated from my grandfather. In the course of time, two of my uncles, Will and

Walter, made good in Argentina; the former by importing pedigree sheep and bulls; the latter married a rich girl and so was able to buy a good estancia (farm) and his younger brothers joined him to help him breed cattle and sheep. Charlie and Frank remained in Argentina during the First World War but returned soon after. They bought a small farm at Wootton, Boars Hill, Oxford, where they lived until they died – Frank in 1967 aged eighty-four, and Charlie in 1973 aged ninety-four.

These two bachelor uncles, so we heard, had at one time quarrelled over a girl who, in the end, turned them both down. This seems to have brought about some bad feeling that remained with them until they died. They never sat together in the same room, Frank preferring the company of their housekeeper, Mrs Small, and her husband. Charlie, on the other hand, sat in the sitting-room; but they took their meals together. Mrs Small even bought Uncle Frank's clothes for him and I can't remember the last time he ventured as far as Oxford. Uncle Charlie had a good business head and ran the farm, giving his brother the orders. They had only one thing in common – cricket. Two other uncles, Jack and Walter, died in Argentina and all the aunts returned to England, after two or three years' stay.

There is an amusing story concerning my Uncle Charlie. His spinster sister, Rose, had come to Middle Farm, Wootton, to end her days. Aunt Rose was taken ill and died in the Abingdon Hospital and the cremation was arranged to take place in Oxford. As neither of my uncles had a car, I offered to pick them up and take them to the crematorium, but the answer I got was, "Jim, we can't go to the funeral the thresher is coming today."

Another of my uncles had no doubt where his priorities lay. This story concerns my Uncle Reg, who had a partnership with his brother Will, exporting pedigree cattle to the Argentine. Another spinster sister had died at Middle Farm and the funeral took place at Wootton Church. While we were all kneeling and the parson was saying a few words, my Uncle Reg, who was sitting next to me, gave me a nudge; I, thinking he wanted a prayer book, took my hands from my face and glanced towards him. He whispered the following: "I had a cable from Will in B.A. this morning saying that he had got a very good price from the sale of the pedigree sheep."

Meanwhile, mother had two more children born in the Argentine; Ethel, born in 1891 and Madge, born in 1894. In 1898 my brother was born, but mother had come back to Boars Hill for his birth. Once more, she returned to the Argentine but things went wrong with the partnership and Recarborda had got in financial difficulties as he spent most of the time in B.A. 'living it up'. My mother was very unhappy in the Argentine and longed to return to England. Certainly the conditions in the 'Camp' were not suitable for her; a lack of English neighbours; and she was unable to speak the native language.

* * *

Father decided he would retire and return to England and rent a farm near Oxford. This, indeed, pleased my mother. In 1900 he set sail from B.A. with my mother, my brother Gilbert, my sisters Ethel and Madge, two horses, 'Captain' and 'Bonny' (for use on the farm) together with a considerable load of Cheddar cheese, which he hoped to sell well in London. However, on unloading the ship and removing the hatches covering the cheeses, he found, to his horror, that the whole consignment had melted, due to the ship passing through the tropics, as tramp steamers did not carry refrigerators. This was certainly a great blow, for the money he had hoped to receive from the cheese was needed to rent a farm.

My family had to live with my mother's people on Boars Hill while my father looked around for a farm to rent. Within a short time he was able to rent a farm called Blagrove, at the foot of Boars Hill, within three miles of Abingdon. It was owned by Lord Berkeley who had recently built himself a castle on Boars Hill. The farm was mostly grass and suitable for dairying and had a small portion for growing cereals. The farm was in very poor condition as there was not a farmhouse, only one small cottage. The original house could be described as a manor and in its time must have been a very fine place. It was sixteenth century, but it had gradually fallen into decay and all that remained in 1900 were the garden walls, which had surrounded the manor, a small stable for the horses, a very large walnut tree and one pear tree of considerable age which was still bearing very small pears when we left the farm in 1933. Blagrove had a lovely oak wood very suitable for rearing pheasants. Lord Berkeley agreed with my father to build a new farmhouse on the foundations of the old manor, a large cowshed for thirty cows, some loose boxes for the horses, a dairy and a cart shed, etc. This was fine and my father was able to make a start, buying a few cows, as my mother had inherited £1,000 from an aunt. The joke was that my mother lent father the money at 2½% interest per year, but from that date till she died in 1948, she never received a penny on the capital that she gave him! When the farm lease was signed, my father rented a house at Old Boars Hill called Rose Bank, which had been, in the eighteenth century, a pub known as 'Black Jack' and there, in 1902, I was born – a 10lb baby.

There is an amusing story concerning my arrival. First of all, my two sisters and my brother, Dick, had gone to stay with their grandparents, the Hedderleys, who lived quite close to Rose Bank. In the house, awaiting my arrival, were my eldest brother, Gilbert, and his great friend, Frank Gillams. Dr Challenor, (locally known as 'Butcher Challenor' because of his partiality for curing most ills with the knife), was in attendance. When the doctor came down from the bedroom, father was told he could go up and see mother. According to mother, all father said was, "Nell, you did kick up a row!" Mother never forgave him for this. Father was not a sensitive man in any way and never appeared to take any interest in his children. I can never remember him ever giving me a kiss during the whole of my life; a very strange man. He never encouraged any of us and left our upbringing entirely to mother.

For some strange reason, my mother's milk was quite unsuitable for me and I could not keep any liquid down. Everything was tried; skimmed milk, Glaxo, etc., but nothing worked. Meanwhile, I rapidly lost weight until I looked like a skinned rabbit. Mother began to think she would lose me and the doctor was unable to suggest any remedy. When nearly all hope was gone, an old nurse heard about me; she called on my mother and suggested that I be fed on whey, the residue from the milk after making butter. Well, mother persuaded father to make butter at Blagrove Farm, and I was fed on warm whey in a feeding bottle. You can imagine the excitement when I kept this fluid down and, what was more, at once began to gain weight. From that day to this I never looked back.

In 1906 the farmhouse and the cowsheds were finished and the whole family moved in. This meant father and Gilbert did not have to walk to and from the farm six times a day.

There are three incidents I must relate concerning my sister Madge, who was given the job of pushing me out in the pram. Rose Bank was situated on a road which was quite steep. It seems, so I have been told, that Madge enjoyed giving my pram a good push so that it was propelled down the slope on its own. Madge, of course, would then dash after the pram and stop it rushing down the hill by grabbing the handle. It was a great game for her – ah! but one day she chased after me in the usual way, but this time she fell down and I went careering down the hill. Fortunately for me, the pram did not keep a straight course and it hit the edge of the road with its front wheels, bounded over the grass to shoot itself and me into a large clump of stinging nettles. My face and hands soon had a rash and I yelled all the way home. Mother took one look at me, found out what had happened, and gave my sister a good hiding.

There was another game Madge played which involved me. She had an old bike with no brakes and with one part of the frame broken, but the use of a broomstick made it rideable. Her great ambition was to be able to ride down Hinksey Hill from Boars Hill to Oxford, at such a speed that she could cross over the railway bridge, which had a steep climb, without further pedalling. This is where I came in for, by putting me on the handlebars and pedalling hard before descending Hinksey Hill, she hoped to achieve her ambition. After many attempts we finally made it. We had to have a clear run, for if we encountered traffic, it was my job to press on the front tyre with my foot and, if I did that, we could never make the bridge. I shudder when I think of it now, with all the traffic that we see on the roads today.

Madge had an adventurous nature and if anyone dared her to do something, she would respond immediately. At Blagrove there was a post and rail fence round the cattle yards about six feet from the ground. Dick and Madge would walk round the yard balancing themselves on the top rail, the idea being to complete the journey without falling off. Any visitors we had were immediately introduced to the post and rails. The most dangerous undertaking was to climb up into the cowshed and walk the length of the shed on a 4 inch beam with a drop of 15 feet on to solid concrete! We took good care that mother and father were not about when doing this.

Of all my family, Madge was the most outstanding character, and not only was she such good fun, but she grew up to be a very good-looking girl. All the games we played at Blagrove, the parties, dances, river picnics, indoor games; it was Madge who arranged them all.

I must relate an amusing incident that happened during a General Election when I was about six years old. My father had two passions, shooting and politics, and he would talk for hours on these two subjects and his adventures in the Argentine. Voting day came and Wootton school was used for the polling station. Father said that if I stood outside the school, I could earn a penny or two by holding the farmers' horses while they went in to vote. This idea was 'just up my street', so father pinned a large blue rosette on my jersey and off I went. It was not long before I had earned several pennies. Suddenly, I noticed that another boy, older and taller than myself had joined me, but he was wearing a large yellow rosette. Not knowing anything about politics, I was astounded when, as I went to take the horse of one of the voters, this boy came up to me, punched me on the nose and knocked me down. I got to my feet, took to my heels and ran home, howling, to my mother, my shirt red with blood. For quite a long time afterwards I would stare at the hoardings with the Tory election posters depicting two cottage loaves – the Tory was a large loaf and the Liberal the small loaf. At that age, I was firmly under the impression that voting was about whether you wanted a large loaf or a small loaf to eat.

There is a sequel to this little story for some years ago I was asked by the local Old Folks' Society to give a talk. They asked me if I would tell them something of the 'Good Old Days'. This I did and ended the talk by describing voting day in Wootton seventy years ago. There was quite a crowd present and sitting in the front row was a man, a little older than myself, who jumped to his feet, exclaiming with laughter, "I am that boy who knocked you down." How everybody roared with laughter!

Just after my first birthday, I was struck down with a mastoid in my right ear which refused to go away. The doctors decided that I must go to the Radcliffe Infirmary and have an operation. In those days, the use of mallet and chisel was the only remedy, so they removed the mastoid bone. This seemed to be all right for a time, but when I was three years old, I had to go back for another operation. My earliest recollections started from that time, for I remember that my head was always bandaged up and crinkly linen 'packing' was pulled out from behind my ear and renewed each day, and this made me cry. I also have clear memories of a blind baby chick and a dog kept in the garden; also going each day for a walk with my mother to a small paddock to hear the skylarks sing. Alas! today the larks have all gone, but the path still remains.

Another memory I have from when I was four years old is of Pat Heavens, who lived next door, an Irish lady of unknown age, whose husband, she always declared, was a sea captain who had run away many years before; but according to village gossip, she had murdered him and then buried him in the garden! My brother Dick, four years older than myself, used to point out the spot where he was buried in the vegetable garden. Certainly the vegetables were very prolific just there! I used to peep through the hedge with a certain amount of apprehension and took good care I was not seen by her.

Pat was a huge fat lady of over 20 stone who wore a fantastic number of petticoats which reached nearly to the ground. She stank to high heaven and her language did the Irish proud. She always had quite a number of little pigs which she reared by hand and which followed her wherever she went. This accounted for the smell in her garden and also from her clothes. My father told me once when he had a long chat with her, the little pigs immediately dived under her skirts for warmth and protection and only came out when she went into her house. Every Wednesday she would harness up her donkey and cart and set off to Oxford to sell her chickens and eggs. It was said she could lift a sack of corn weighing 2½ hundred-weight! Part of her house alongside the road had walls covered with corrugated tin and if you wanted to hear her language, all you had to do was to run past the fence rattling it with a stick in your hand, thus making a terrible noise. It never failed to fetch her out, shouting abuse at the children as they disappeared. All children were scared stiff of her.

Blagrove Farm had been neglected for many years. The hedges were like concertinas, the arable land was full of 'squitch' and docks, and the grassland in many places had rushes in the furrows. But, oh boy! it was a wonderful place for us children to be brought up in. There was so much wildlife to be found in the hedgerows and in the oak woods; nuts, bluebells, primroses and many other wild flowers, and the whole farm teemed with rabbits, hares, partridges and pheasants.

The old manor, or what remained of it, was reputed to have the ghosts of four white horses and a postilion who, at midnight on winter nights, drove in a carriage up the drive, which was lined with huge elm trees. Then there was the story of hidden treasure from the old Benedictine Abbey of Abingdon, which was supposed to be buried at Blagrove, consisting of silver chalices, crosses, plate, etc., hidden from Henry VIII by the monks. We children had phases of furious diggings, but nothing turned up except pieces of late Victorian china and these were found in the stump of an old oak tree, some distance from the farmhouse. How they got there, we never knew.

By 1910 my eldest brother Gilbert could not agree with my father on the running of the farm, so he decided to go to Canada and we did not see him again until the 1914/18 war, when he was a soldier in the famous

Canadian regiment of the Princess Patricia Canadian Infantry. He survived the war and returned to Canada, married a girl from the Argentine and then went to New Zealand in 1927.

I now lead up to one of the most important events of my life – my first visit to Youlbury and my first meeting with Sir Arthur Evans.

Even after the two operations on my ear, it was still discharging. At one point, I had to go to a doctor to have it syringed with what to me looked like a garden syringe, but with little result. Then followed drops of various kinds which never really helped. (It was only after I had lived in the Argentine that the discharge stopped, but even then, from time to time, it would start up again. Now I have very little trouble with it, but I still go to the Radcliffe Infirmary for a check up and each time I go, the young doctors are amazed to see my ear and call one another to have a look at how they operated on mastoids in those days.)

I was five years old when my mother took me up to Youlbury for the flower show which was held in a paddock belonging to Sir Arthur Evans. There were all the usual side shows, stalls, etc. but the most important event of the day was the tug-of-war between the villages of Wootton and Boars Hill, together with Sunningwell. There had always been great rivalry between these villages; we thought that Sunningwell people were almost 'foreigners', and as for a Wootton boy marrying a Sunningwell girl, or vice versa, eyebrows were raised very high. At last the moment arrived for which all had been waiting – the tug-of-war. My mother was holding my hand, we were at the back of the crowd and I could not see what was happening for a sea of legs. At that moment, a short gentleman came up to my mother and asked if the little boy would like to have a better view of the tug-of-war. Mother replied "Certainly." I was then hoisted on to his shoulders and I had a grandstand view of the event. I don't remember who won. My mother addressed him as Dr Evans, as he was then, for she had previously met him in her sister Ethel's house, where he was staying while his house was being built. He then suggested to my mother that perhaps I would like to see the lake he was making in an old gravel pit, so off we went. I was delighted to see, near the edge of the lake, a railway track and little wagons loaded with clay, but I was a little disappointed to learn that there was no engine, as I had a toy engine at home. The clay was for puddling the lake to keep the water from running out. The lake, when completed, would be close on four acres and in some parts forty feet deep.

* * *

No dog has yet surpassed Blagrove Jock, known for miles around Sunningwell, Wootton and Boars Hill, as the best rabbiter and rat catcher. We won't go into his pedigree, for he had none, but let us just say that he was a friend to everyone. My brother, Martin, learnt to walk with the aid of Jock and he loved to dress him up in a pair of old red shorts that had a hole for his tail to go through, and a green jacket, buttoned up the chest, with the arms cut out. Poor Jock suffered this indignity, but the final straw was being made to ride in an old pram. What a 'hang-dog' look he always wore when being pushed around in it!

One day, Martin was pushing him round the tennis court, when a rabbit jumped up from his squat in the grass and bolted for his warren across the field into a large thick bramble bush. This was just too much for Jock and he leapt from the pram and, in full cry, gave chase after the rabbit. It was a lovely sight to see a dog dressed in red shorts and green jacket chasing a rabbit across the field. Needless to say, the rabbit reached the bramble bush and disappeared down its hole, but poor Jock became completely entangled in the brambles and we had to cut him free. After this episode, Martin was never allowed to dress him up again.

All my life, I have been interested in shooting game, an interest that I think I inherited from my father. His greatest hobby, apart from shooting game, was rifle shooting. Every year he would go off to Bisley for two weeks in June and even at eighty years old, was still shooting for the King's Hundred. I suppose I was about twelve years old when at last I succumbed to the temptation and took father's twelve bore gun while he and mother were away at Oxford Market. When Jock saw me reach for the gun, he was beside himself with excitement and dashed off to the nearest hedge. It was not long before he found a rabbit, but it kept running down the side of the hedge with Jock close on its heels. Several times I saw the rabbit but I was afraid to shoot it in case I hit the dog. At the bottom of the field, there was another hedge at right angles, with a ditch in front. The rabbit decided to cut the corner and make for the other hedge, leaving Jock in the old one. At last! I lifted the gun and fired at the rabbit, but missed. To my absolute horror, a tremendous shout went up and there, from the ditch, appeared our handy man, Teddy Hutt. I had no idea that he was cleaning out the ditch as clumps of nettles had obscured him. The shots from my gun hit him in the backside but, thank goodness, he was beyond the range of receiving any serious damage and they had only stung him. Fortunately, his cord trousers had kept out the shot. Naturally, he was very angry and he gave me a good dressing down, but I persuaded him not to tell my father by giving him my week's pocket money of sixpence. Teddy kept his word and I never touched a gun again until I was seventeen years old. On the last day that Teddy worked for my father, I reminded him of what had taken place all those years ago. "I never told your dad because I knew what it cost you to give me that sixpence," he said. I can see him now on his bicycle, pedalling with his heels because his feet were malformed. It was wonderful to walk behind him on a dewy morning, for his feet were turned out at nearly ninety degrees.

One warm September day in 1912, I was with my brother, Dick, Lawrence Greatbatch, a friend from Oxford, and our mongrel dog Jock. As I have already said, Jock was one of the best dogs I have ever seen for catching rabbits, and as my father never thought to give us any pocket money, we had to depend on Jock to catch rabbits – and with the aid of a spade we were very successful. All the rabbits that we caught we took to Sunningwell to Mrs Silvester who kept a little shop in her house, with a bell behind the door to warn her of any customers. With the proceeds from the rabbits, usually sixpence each, we bought ourselves chocolates and ginger beer, etc., and always a packet of chocolates for Jock. On this particular day, we had managed to kill three rabbits, and as all of us were tired out by our exertions, especially Jock, we soon made ourselves comfortable under a lovely large spreading oak tree. As we were talking, Lawrence Greatbatch suddenly made a startling suggestion; why did we not form a troop of Boy Scouts? (He was, himself, a member of the first troop to be formed in Oxford.) We declared it a splendid idea and immediately contacted the two sons of Lord Berkeley's gamekeeper, and two boys working for my father, who agreed to join us. We then sent for Lord Baden Powell's book on Scouting and we used my father's cowshed for a temporary headquarters. We had enough money from killing rabbits to buy the old type of Scout's hats and persuaded mother to part with her old broomsticks. The first thing we did was to learn our knots and the Scout's law. Meanwhile some more boys from Wootton village had joined in, making us twelve in all. Then we found out that we were illegal because we had never taken the Oath to serve God and the King, which we would have to do before a magistrate. So we all marched down and presented ourselves in front of Mr Preston, the magistrate, who took the oath from each of us and this very much impressed us. He asked if we had a Scoutmaster and on hearing that we had not, advised us to get in touch with Arthur Shepherd who lived on Boars Hill. The next day the three of us approached him and asked him to be our Scoutmaster. He was filled with enthusiasm for the new troop and at once wrote off to London to the Scout HQ for all the literature on the subject. After a few weeks he got his Warrant and that made us legal.

The next thing to be done was to find a suitable place where we could do our training without antagonizing the local farmers, who were suspicious of small boys dashing about their farms, leaving gates open, making fires, etc. We found quite a good place on the south slope of Boars Hill, below Lord Berkeley's fenced-in woods. This area consisted mostly of gorse bushes and rough grass. What a wonderful time we had that spring and summer, learning our knots, etc., and earning our badges, which all Scouts like to do. We did not give up our hunt for rabbits so that we could cook them in the evening over camp fires. The one important item we lacked, was a Scouts' HQ so that we could carry out our scouting in the winter months. Arthur Shepherd was an excellent Scoutmaster and filled us with great enthusiasm and it was he who started to raise money to rent a hut for our use. Then, one day, when we were gathered among our furze bushes, he informed us that we were on land belonging to Lord Berkeley and that we were disturbing the pheasants in his wood when we went gathering wood for our nightly bonfire for the evening meal of stewed rabbit. In fact, Arthur Shepherd was told that we had to leave at the end of the week. I don't suppose that Lord Berkeley knew anything about it; I suspect it was the keeper's orders. You can imagine how we felt at that bombshell, for we thought that we would have to give up scouting altogether.

While sitting round our camp fire, Arthur Shepherd suddenly had a bright idea. He told us he had thought of a possible solution to our problem. He would go and see Sir Arthur Evans who had just returned from Crete where he had discovered evidence of the wonderful Minoan Civilization and the Palace of King Minos. He jumped on his bike and rode off to Youlbury, the house of Sir Arthur, leaving us very despondent, sitting round the fire, wondering what would be the outcome. About an hour later we saw him pedalling like mad towards us, waving his hands and crying, "Cheer up! Come on boys, clean your shoes, wash your hands and faces, put on your hats, tighten your knots and line up, for we have to march up to Youlbury and be inspected by Sir Arthur."

You can imagine our faces when we pressed the bell and there stood Sir Arthur, waiting to receive us, standing in his stately hall which was laid out in black and white marble in the form of a labyrinth with the Minotaur in the middle. What caught our eyes most was the spread on a trestle table – buns, cakes, biscuits, sandwiches, tea and coffee. Sir Arthur told us that he was delighted to see us and that the Scout movement was one that he would back up to the hilt. Furthermore, he said that Youlbury wood of seventy acres was there for our scouting, then, smiling, he added, "I don't preserve pheasants and you may kill as many rabbits as you like." How we cheered and clapped. He then said he would provide, in the woods, an HQ for our sole use and, as well, we would be allowed to swim in the lake two evenings a week, provided we could pass the swimming test, and that Arthur Shepherd would be present when we were bathing. All this, when the news got around Wootton, Boars Hill and Sunningwell, brought in a lot of boys wanting to join.

What a wonderful HQ Sir Arthur gave us. It was built in the form of a pile-house, i.e. by cutting off the tops of fir trees and building a floor firmly attached to the sawn-off trees, thus giving a base on which to build. A wide veranda went round the summer house and it also had steps up to the platform, which was eight feet off the ground, giving wide views over the lake and the Berkshire Downs. This summer house had been built for his wife who had suffered from TB but she died before it was finished. What a splendid Club House it was for us all. Not only did we use it for scouting nights, but on certain occasions we slept there, for it was fully equipped. I often think of the stewed rabbit suppers we ate, thanks to my dog Jock. Our first summer camp was in 1912 and Sir Arthur provided four bell tents for us to sleep in which were pitched in a field close by.

On scouting nights Sir Arthur used to come and visit us. He was so friendly and would watch us do our exercises and the preparation for winning our badges and all the different games that Scouts so enjoy. He also provided transport when we went to any rallies in other parts of Oxfordshire and Berkshire. At this point, my brother Dick left the Scouts, because he went to France to join an engineering firm in Lille. Lawrence Greatbatch also left and went into his father's business. This left me the only founder member. The Scout Troop was known as the 29th North Berks and a flag was given to us by the local Commissioner of Scouts.

We were quite overawed by the presence of Sir Arthur when he visited us on scouting nights for, at about this time, the whole world was excited by his furthur discoveries in Crete of the Minoan Civilization, and we were especially excited by the legends of King Minos, his daughter Ariadne, Theseus and the Minotaur.

Looking back on those days, I gradually became aware that he liked to talk to me and to join in certain games, such as 'Flag Raiding', which entailed crawling on one's hands and knees through the bracken in order to capture the opponents' flag, without being seen by them. It seems that Sir Arthur had noticed that I was very pale and he asked Arthur Shepherd if he knew the reason. He told him that I was suffering from an infected mastoid and that after two operations there had been no improvement, and that this was affecting my health, as the ear discharged continually.

It was a Saturday afternoon and we were all at the HQ when Sir Arthur arrived and asked me if I would like to have tea with him and walk round the gardens and the lake. To me, his house was out of this world. The luxury, the elegance, the vastness of the rooms, the pictures, the tapestries and the huge library with books of every description; but what impressed me most at that time, was the number of bedrooms, twenty-two in all, the five bathrooms and, above all, a Roman bath with three steps going down into it. What fun I had later with my school friends, for it was possible to do two breast strokes across the bath. (One day when we were fooling about, the water overflowed and went out into the passage.) And to think that Sir Arthur, with all these bathrooms, preferred to use, all his life, a large tin bath which was kept under his bed. On the landing outside the bathroom, was the head and shoulders of a large brown bear, which had attacked Sir Arthur's brother, Norman, when he was in the Carpathian Mountains. It was so arranged, in a large glass cage, that the bear appeared to be coming out of its lair; the foliage and rocks made it look so authentic. The house was a regular museum of Cretan discoveries and fine collections in glass cases of Stone Age implements and New Zealand jade ceremonial weapons.

After this particular visit, so vivid in my mind, I became Jimmie to him until he died in 1941.

* * *

Unknown to me, Sir Arthur made many visits to Blagrove to discuss my future life with my father and mother. The outcome of these meetings with my parents was that, first and foremost, Sir Arthur wanted to send me to the best ear specialist in London, and also to be responsible for my schooling so that I could have the best education. Everything, from now onwards, would be paid for by him – in other words, he would be my guardian. This was a momentous decision for my parents to have to make, but after due consideration, they decided that this was a wonderful chance for me and agreed with Sir Arthur's generous offer. My mother was distressed at parting from me, but she felt she could not stand in the way of such a wonderful opportunity. I would be able to see my family every Sunday after church and when I came back from school I would be able to spend the first ten days of my holiday with them; also a few days before returning to school.

The great day arrived, at last, when Sir Arthur sent his car to Blagrove to pick me up. There was a mixture of tears and smiles on my mother's face as she kissed me and I, too, felt sad because I loved Blagrove and my brothers and sisters.

It was a case of farmyard to luxury and when the chauffeur rang the bell and Emma, the parlourmaid, took me to Sir Arthur with the words, "Sir Arthur – Master Jimmie," my whole life was to change. The first surprise was that Sir Arthur kissed me. This was something I did not expect as my father had never kissed me in my life. He then took me upstairs and introduced me to Ada Porter, the housemaid, who was to have charge of me. Oh dear, what a wonderful mother she became to me. I loved her from that day until she died. I was so very sad on hearing of her death from Sir Arthur, when I was in the Argentine. For any problems I had, or if I was homesick, which I was at first, there was always Ada to comfort me. I had a lovely bedroom next to Sir Arthur's, overlooking the garden and the lake to a glorious view of the Berkshire Downs. He had a bell-push installed which rang in his room so that if I needed anything in the night he could come in to comfort me.

Next day, Ada took me by car to Oxford to buy a complete new set of clothes. I had never seen such an assortment of clothes that I took back to Youlbury, including knickerbockers, stockings, an overcoat, a suit for Sundays and an Eton jacket, (for I had to change every night for dinner), and all my shoes were to be made for me. What a change from my usual hand-me-downs. All my old clothes were sent back to Blagrove and, no doubt, came in handy for my brother, Martin, who was four years younger than I.

At 8 o'clock in the morning Ada used to come into my room with a plate of fruit; peaches and nectarines in the summer, grapes, apples and plums during the autumn and winter months; all grown at Youlbury. All my clothes were put out for me for that day, and a can of hot water with a towel over it, was placed in my basin. At 9 a.m. the gong went for breakfast. What a spread! Every day, summer and winter, on the sideboard there was porridge, cold ham or kippers, eggs and bacon and kidney, all kept hot, and we would help ourselves to anything we wanted. On the table were boiled eggs, coffee, toast, marmalade or quince jam. At lunch-time it was quite different, for we had two maids to wait at table, Emma, Sir Arthur's personal maid, and her assistant, Maud. On the table, in front of each person, was a white menu. There were always three courses, followed by cheese and fruit. At a given signal, Emma would open the door, Sir Arthur would leave the table and we would follow him to South Hall for delicious coffee served in Crown Derby cups.

After lunch, Sir Arthur always had a short siesta in the library, and then at 3 o'clock, we would go for a walk. This was followed by tea in the drawing-room, which was the most magnificent room in the house. Emma would bring in the tea things and put them in front of Sir Arthur, who would be sitting in his favourite chair, and he would serve the tea. There were always sandwiches, tomato and cucumber, followed by two types of cake. In the summer, if the weather was fine, we had tea on the terrace. At 6 o'clock, guests or no guests, Sir Arthur would leave the room and go to the library to write letters.

Dinner was at 7 o'clock sharp and woe betide you if you were late. At 6.45 the gong for changing would go and I had to change into an Eton jacket and be downstairs just a few moments before 7 o'clock. Ada, of course, at 6.30 would have put more hot water in the basin, made up my fire and laid out all my clothes for me to change into. When the final gong went for dinner we would go, headed by Sir Arthur, dressed in a dark suit, to the dining-room where Emma was standing by the door which was then closed by her. The table was large and round, beautifully 'dressed' with flowers, silver, etc., two types of wine glasses and smaller ones for port. There were never any pre-dinner drinks, but all wines were of the very best and, if guests were present, champagne was served. A typical menu was oysters, pheasant with all the trimmings, pudding and a savoury, followed by any fruit which was in season from the glasshouses. Coffee was never taken at the table, but always in the drawing-room. While we were having dinner, the maid would go to the drawing-room, rearrange the cushions, etc. and in winter make up the fire with pine wood logs and a handful of dried rose petals. What a lovely smell to greet you when you came in to sit down and drink your excellent coffee.

At 9 o'clock Emma would arrive and put a decanter of whisky and a beautiful silver box containing biscuits on the side table, and at 9.30 Sir Arthur would retire to bed, leaving any guests who were staying to amuse themselves. But, of course, I had to go too, for Ada was waiting to give me my bath and tuck me up with a kiss. Later on, when I was older, I could stay up with my friends, but Sir Arthur never changed *his* habits.

Now I think the time has come to relate what this extraordinary man was like. When I first went to Youlbury in 1912, Sir Arthur was sixty-two, and he seemed to me to be the kindest man that I had ever met and I never had reason to change my first impression of him. He was at his best among children and it was a pure joy when he was in their company, for he came down to their level. He knew how to talk to them and had a lovely sense of humour. Kindness poured out of him; but, of course, when he told you to do something, by Jove, you had to comply. In that way he was a Victorian. He was quick to anger, but just as quick to forgive, stubborn, and considered by some of his archaeologist colleagues, to be a bit of a tyrant. Many a time at an hotel, Sir Arthur, after giving an order for lunch or asking for a knife or fork, and the waiter kept him waiting, would 'blow his top' and would tick him off left, right and centre, with everyone watching. Many a time I wanted to sink through the floor. The first sign of his getting angry was when he started scratching the back of his head; it reminded me of my grandfather rubbing his thumb nail up and down his waistcoat buttons.

Now back again to my arrival at Youlbury. The first thing Sir Arthur did was to ask Miss Mary Wiggins, daughter of the Vicar of Watlington, to come back to Youlbury to assess my academic ability and to be my governess during the holidays. She had earlier been engaged when Sir Arthur adopted his nephew, Lance Freeman, when he was quite young. When I arrived at Youlbury, Lance had finished his education at Harrow and was training to be an officer at Sandhurst.

It soon became apparent that my education left a lot to be desired. It had suffered through my health and I had been at the Abingdon Convent for two terms and then withdrawn when my parents, due to financial difficulties, decided to start a little school at Blagrove. The school consisted of my two Candy cousins, three cousins from the Hughes family, who were farmers at Wootton village. A young woman was engaged to teach us, helped by my sister Madge. It was an all day school, with my mother providing the midday meal, for which she was paid. That sort of education would not be allowed today.

Sir Arthur and Miss Wiggins decided that a small boarding-school, where I would get individual teaching, would be the best thing for me. They chose a school on the Norfolk coast at a place called Snettisham, noted for its bracing climate, ideal for me; also the school specialized in 'backward' boys. I shall speak more of this later.

In the meantime, as my ear did not improve, Sir Arthur took me to London to see a specialist in Harley Street, who advised that my tonsils should be removed. This was done, but still my ear discharged. I might add that throughout my life, I never suffered with earache.

Meanwhile, what a contrast I found comparing life at Youlbury with that of Blagrove; to find myself surrounded by priceless things collected not only from Crete and the Balkans, where Sir Arthur had been travelling after leaving Oxford, but also from many other places. What fascinated me was his wonderful collection of gold and silver coins, partly inherited from his father and added to by himself. I never ceased to wonder at the fantastic skill of the engravers of the seal stones from Crete and, above all, the famous collection of silver coins from Syracuse on which were engraved such things as goddesses riding on dolphins; every hair on their heads you felt you could pick up. All these precious things were locked away. Then again, there were beautiful pictures to look at. But best of all, I loved the wonderful stories that he told me of his journeys through the Balkans, Europe, Finland, Greece and, of course, Crete. His library had a fantastic collection of learned books, many in foreign languages.

Of all the servants, Ada was my favourite. Mrs Judd was the housekeeper and cook, and sometimes, I used to try and get her to change the pudding for lunch or dinner before she went into the library with her slate containing the menus for that day, for approval by Sir Arthur. In her department was one assistant cook and a scullery maid, also a boy who cleaned the shoes, fetched in the wood for the central heating (1913) and attended to coal supplies, etc. First thing in the morning, all the dusting and cleaning was done by the downstairs maids, four in all, who were dressed in morning dresses which they changed after 12 o'clock for black dresses with white aprons and caps. No silk stockings! And while we were having breakfast, the bedrooms were done by Ada and Edie.

Outside, there was the head gardener, Mr Osbourne, who had four men under him, looking after the greenhouses, cutting the lawns, attending to paths, flowerbeds, etc. and the conservatory. The lake, the woods and their paths, the poultry and milk for the house were looked after by two men living in the lodge by the gates. Then there was Jim Wiblin, the head chauffeur, who also acted as Bailee to the whole estate, except the gardens. Under him was Charlie Mott, who did nearly all the chauffeuring and looked after the electricity generator.

But, at times, I longed for Blagrove, for my mother, for Jock, the rabbits, bird nesting and wandering about the farm and fields of Shivering Dick, moon daisies, and cowslips which I used to pick for my mother.

At last the day came for me to go to school at Snettisham, which is not far from Sandringham in Norfolk. This was the autumn term in 1913 and I was eleven years old. I travelled in the guard's van, the guard having been instructed to see that I got out at King's Lynn. Sir Arthur kissed me when I left London and that really touched me. I must, at this stage, quote part of a letter written to my mother, which she handed to me many years ago, along with others, outlining the idea that Sir Arthur wished to be my guardian:

You know that I have really given Jimmie a little bit of my heart.

I think I have twenty-seven letters addressed to my mother and over a hundred-and-twenty to me from Sir Arthur covering the period from 1914 to 1941. In the early days, Sir Arthur always started his letters to me with 'My dear Kid'. Greece and Crete contained many goats and Sir Arthur always contended that I would grow little horns, so on my arrival back from Snettisham for the holidays, he used to feel my forehead and insist that the horns were sprouting.

Later on, he was to start his letters with 'My dear Jimmie'. These letters and the envelopes were always edged in black and on the back of the envelope, appeared the black head of an elephant. He used these black-edged letters and envelopes in mourning for his wife, Margaret, who was the daughter of Professor Freeman, and who had died in 1893. All his correspondence and even the three volumes of his *Palace of Knossos* were written with a white goosefeather quill pen.

* * *

Snettisham School was run by two brothers, called Prosser, both bachelors, to prepare boys for entrance into public schools. They had a most devastating effect on me, for certain, and I should think on the other boys as well. The elder brother, whom I feared most, taught every subject except music, which his younger brother taught. As I have already stated, I had very little basic education due to my health and lack of instruction. There were twelve of us so-called backward children, and my weaknesses were spelling, maths, especially algebra, and Latin. I was all right in history, geography and the Bible. Mr Prosser would come up behind us and lean over to inspect our work, holding a green pencil in his hand. If I did not know the answer to his question, I would, in desperation, say or write down anything. That would set him off and I became like a rabbit with a stoat ready to pounce – in other words, petrified. He would then put his green pencil between his teeth, lift my jaw up with his left hand and smack with his right hand the taut flesh of my right cheek – very painful. This was called 'cracking' by us boys. We all got it at some time or other but I and another boy suffered the most. What we did not know was that Mr Prosser suffered from ulcers which was probably the cause of his harsh treatment of us. We were very superstitious about the colour of his tie. In the mornings we would wait at the bottom of the stairs for his arrival before going into breakfast, and if we saw he was wearing a red tie, we then knew we were in for trouble during lessons, but if, on the other hand, he was wearing a green one, then the day would go well for us; very strange, but it was true.

Sometimes there were terrible scenes over maths, and I don't know how many times I had to write 'I must not guess'. His favourite saying, when he was rowing me, was, "You are the personification of a lazy lout," or that I was "an unmitigated fool." Things went from bad to worse, and, on one occasion, he beat me with a broken golf club across my back. At that time, I thought all boys went through that sort of treatment, having read the books of Dickens. In my unhappy state I became very religious and during break, I used to shut myself in the loo and after I had had a bad period in Latin or maths, pray to God. When the other boys saw my back in the bath, they took a photograph and suggested that I told my mother. This I would have done, but I did not want her rushing to Sir Arthur, so I decided to write to Miss Wiggins instead. The result was that Sir Arthur sent a telegram to Mr Prosser asking that I might come to London to be interviewed for a public school. The next day I set off and Sir Arthur was waiting for me at St. Pancras station. I never returned to Snettisham School and in due course my trunk and bicycle were sent on to me. I did hear that after the term finished, the school did not open again.

Sir Arthur made a strange prophecy while we were at St. Pancras station. It was at that time in the Great War when zeppelins had been seen over the Baltic and it was well known, even before the war, that the Germans had them. When Sir Arthur kissed me, just before the train pulled out, he said to me, "Mind, Jimmie, the zeps. will come tonight and drop a bomb. So look out!" Sure enough, while we were all fast asleep in our beds, a zep. did come over the village and dropped one small bomb. It just missed the church, but blew out all the windows. The noise did waken me, but I soon fell asleep again. Next morning, we all went to see the big hole, in a field, close to the west window of the church. Another bomb was dropped on the village of Heacham, a few miles away. It fell on the straw roof of a cottage and rolled into the water butt and it was only discovered when the old woman living in the cottage climbed up to look inside to see what was blocking the outlet at the bottom of the butt! To her horror, she saw the bomb and called the police. It was then removed and blown up. In those early days, they were very small bombs and were thrown out of the zep. by hand. Sir Arthur was delighted that his prophecy had come true, and told me of a previous occasion.

This had been at Knossos in Crete when he first started excavating in 1900. To start digging with a large gang of men, it was essential to find a good supply of water. He looked around and decided on a certain spot and told his foreman, a man who had lived all his life in that particular district, to start digging. The foreman argued that it was the last place to look and indicated an area somewhere else. Sir Arthur insisted that his spot would be the only place where a well would be found. The foreman replied, "It's your money so we will dig where you want, but you will be wasting both your money and your time because it is not there. I can guarantee that." After a very short time, they discovered a Minoan well with a good spring and, at the bottom, a pitcher two thousand years old. After that, Sir Arthur's workmen thought he had divine powers.

After my leaving Snettisham Grange, Sir Arthur had to look for a public school that would take me without an entrance examination and, in the meantime, I stayed at Youlbury, with Miss Wiggins giving me lessons each day. In time, and after discussions with my parents, he found that Trent

College would take me in the September of 1916.

* * *

I must now retrace my steps to the summer of 1914 and talk again about our Scout camp at Youlbury, where we were joined by the 2nd Oxford Troop. Our Scoutmaster was Frank Gillams of the 29th North Berks., a very great friend of mine and all my family, until he died, aged ninety. He was, indeed, a wonderful man to all Scouts and was a born Scoutmaster. Mother used to say of him, "If the children are with Frank then I need not worry." He took over our Scouts when Arthur Shepperd was killed in his plane in 1913 – a sad loss indeed. He and Sir Arthur saw eye to eye on all Scout matters and he had free access to Sir Arthur at any time. Frank taught us all to swim in the lake, to box and to play football, for he was goalkeeper for Oxford City. He was a boat builder for Salter Bros. of Oxford and spent his whole life with that firm and he was a prodigious walker, sometimes walking as much as thirty miles a day.

At the camp at Youlbury, in that fateful August 1914, we had quite a large gathering of Scouts from Oxford as well as some older boys who helped out. We were all in bell tents and I had my faithful dog, Jock, with me, who still kept us in rabbits! One night Lance Freeman, Sir Arthur's nephew, and three of his Sandhurst friends, on leave and staying at Youlbury, raided our camp and let down the guy ropes, causing confusion and great amusement amongst us all. My tent was not touched at all; Jock saw to that! Every day Sir Arthur would arrive with his car filled with all sorts of goodies; fruit and vegetables from his garden to supplement our rations. What kindness!

The war was very much on our minds as nothing like this had happened since the Boer War. August the 4th arrived and Sir Arthur came on foot and asked Frank to call all of us boys together, as he had an important announcement to make. When we had all gathered round in a circle, he told us that Great Britain was at war with Germany. How we shouted "Hooray, hooray," and cheered and cheered. How could we young ones know what war was really like? But we soon learnt, when one or two of the older boys, who helped Frank to run the camp, had died in Flanders. It was sad to see, as the war continued, our older Scouts joining the forces and most of them never returned.

Just before the war, Sir Arthur had built a large wooden tower near the site of the old cairn, in order to see the wonderful view over Oxford, the Chilterns and the Berkshire Downs. In some way, it was a modern type of folly. I think the tower was 150 feet higher than the highest point on Boars Hill. It had a platform and veranda at three different levels. On the final platform where only the bravest could look down with comfort, Sir Arthur, on the outbreak of the Great War, erected four flag poles with the idea that should a victory be claimed by one of the Allies, their national flag would be flown. This would enable the local residents to be kept informed on the progress of the war and which of the Allies was responsible. Now it happened that Sir Arthur, over many years, was very friendly with Seaton Watson, editor of the *Manchester Guardian*. At 11 o'clock in the morning, Sir Arthur would telephone Seaton Watson for information on the progress of the war. (One must remember that at that time there was no wireless and the public had to wait for the next day's papers to read what was happening.) This information obtained from London was of importance, for any allied victory on land or sea was known at Youlbury long before the general public could see it in the national papers. My father at Blagrove, with the aid of his binoculars, could see what flag was being flown from the look-out. Sad to say, sometimes no flags were seen for some considerable time. After many years of service and constant use by the public, the look-out became unsafe and was taken down.

Many years later, Sir Arthur, at his own expense, built Jarn Mound, a sort of miniature Silbury Hill, and a wild garden on Boars Hill, so that the public would still enjoy the views over Oxford and the surrounding country.

* * *

Let us now return to 1913 again and relate an amusing incident concerning an uncle, by marriage, Jim Hughes, a local farmer living at the Manor Farm, Wootton. One morning he was not feeling very well and so stayed in bed. Uncle Jim was scared of being ill in any way and as he lay moaning in bed, my aunt decided to take his temperature and, to her horror, it was 104°. The effect of this on my uncle and aunt was beyond description as my uncle decided at once that he was on his 'death bed'. He summoned the parson, the doctor and his solicitor and ordered the children back from school, the four girls being with me at Blagrove School. My mother decided to rush to her sister at Wootton to help in the crisis. On arriving, she found her sister Loui in a flood of tears and Uncle Jim moaning and groaning and firmly believing that he was about to meet his Maker. My mother took the thermometer, shook it well down and popped it into my uncle's mouth. The four girls were crying their eyes out, and everyone else was holding their breath. When my mother removed the thermometer, it read just over normal. No one would believe the reading, so mother repeated the operation and the result was the same. My uncle was the first to react; he jumped out of bed demanding his trousers and his lunch! It transpired that a few days previously one of the children had had a temperature of 104° and nobody had shaken the thermometer down.

I must once again digress and bring in another character, my Great-Uncle Will Candy of Winchester. It always seemed strange to me, when I was a little boy, to hear my father call him 'Uncle'. This great-uncle of mine, born in 1839, was a farmer and a cattle dealer. He had two sons and a daughter and I can just remember staying one night at his farm with my father when I was eight years old. From all accounts, he had a bad temper first thing in the morning, but as soon as he had had his breakfast, he was perfectly normal. What used to happen occasionally was that he would get up and go out to the farmyard adjoining the house to tell the men he employed milking, ploughing, stockraising, etc. their duties for the day. On arriving in the yard, he was sure to find something he did not like; either someone who had arrived a few moments late or a job left undone from the previous day, and in an instant he would 'blow his top', sack the man who had arrived late or the man who was responsible for the undone job. If he saw the slightest smile on any of the other men's faces, they were promptly sacked as well. The boys would, on these occasions, hide behind the men, but woe betide them if he saw them with handkerchiefs stuffed in their mouths to stop their laughter! Within a few minutes he had generally sacked the lot and then he went in to have his breakfast. One would surmise that his staff would go home, but not a bit of it; they all knew that he did not mean a word he said and they were so used to these outbursts and, in fact, they liked him. My poor aunt was 'for it' if breakfast was not ready when he went into the house after one of these outbursts of temper. This particular trait runs in my family as well. My mother always said the only time she ever had words with my father was before breakfast. My brother Gilbert was just as fiery as my Great-Uncle Will, but it quickly passed as soon as he had eaten breakfast. I think the reason for these bad tempers was having to get up at 5 o'clock in the morning to milk the cows, after just one cup of tea, then returning to the house to find no breakfast ready and, even worse, the fire had not been lit. Even my dear daughter, Jenny, is very quiet until she has eaten her breakfast.

When my brother Gilbert arrived back in England as a soldier in 1914, his regiment was training at Winchester, so he had a chance to spend a few hours with Great-Uncle Will. Their training, at that time, seemed to consist of marching in columns of four, up and down the old Roman road around Winchester, the idea being to harden their feet for marching long distances in France. One afternoon, on returning from a route march of thirty miles, who should they see on approaching Winchester but my Great-Uncle Will in his dog cart and high-stepping horse, setting out to buy some cattle. The normal procedure was for civilian traffic to pull to the side of the road to allow the soldiers to pass in formation. When the officer saw my great-uncle driving at his usual pace towards him, he shouted to him, "You get off the road and pull into the side." This order of course had the opposite effect, and my great-uncle at once reacted by getting out his whip, flicking his horse to increase its speed and driving right through the ranks of the soldiers, who naturally scattered to the side, and then continuing on his journey. When Gilbert saw who it was in the dog cart he kept his head well down. At supper that night, Great-Uncle Will related what had happened in the afternoon with the soldiers. His remarks were, "I'm not going to be stopped by any God-damned army, and I have every right to travel down any public road in the country." Gilbert thought it advisable not to say that he was with the column of men marching back to barracks.

Now back to Sir Arthur and Youlbury. I don't, for one minute, wish to describe Sir Arthur's archaeological excavations at Knossos on Crete or his very extensive travels through Bosnia and Hersegovina, as a young man. All these accounts can be found in the biography by his half-sister, Dame Joan Evans, in *Time and Chance*; also, recently published, Sylvia Howitz's book *Find of a Lifetime.* I shall content myself with continuing to describe, in some detail, this extraordinary and lovable man who had such an effect on my life.

After returning from Snettisham for the holidays and spending a week or ten days with my family at Blagrove, Sir Arthur would send the car to fetch me to Youlbury. Miss Wiggins would be already ensconsed there waiting to give me lessons the next morning. As soon as breakfast was over, Miss Wiggins and I would go to the morning room where she would teach me English, spelling, maths, etc. This would continue until midday, when Sir Arthur would take me for a walk in the gardens and round the lake. I thoroughly enjoyed this little break for we always visited the walled fruit gardens and glasshouses to help ourselves to peaches, nectarines and grapes. The peaches and nectarines attached to the walls had, tied under their lower branches, a muslin net to catch the fruit when it fell which meant, of course, that it was perfectly ripe for eating. In the centre of the walled gardens was a very large fruit cage eight feet high, which protected the strawberries, gooseberries, raspberries, etc. from the birds. I was not allowed to stuff myself with fruit whenever I liked, as Mr Osborne the head gardener saw to that. (One thing I was allowed – in fact encouraged to do – was, after bathing in the lake, to eat as many gooseberries as I liked.) Then we would continue our walk round the lake, up through the beautiful gardens to the house for lunch at 1 o'clock. In the afternoons I was free to do what I liked. In the summer, I spent a good deal of my time in the canoe or reading on any of the three islands. After lunch, while we were drinking our coffee, Emma always asked Sir Arthur if there were any orders for the car that afternoon. If the weather was fine during the summer or autumn, he would order the car for 3 o'clock and off we would set with Miss Wiggins to visit places of historical interest such as churches, castles, stately homes and, above all, ancient stone circles like the Roll-Right Stones, Avebury, Stonehenge and others. He would tell me the origins and the folklore which had grown up for centuries around them. If he found a church door locked, which was not very often, he would soon find the person who had the key. Then followed tea with plenty of bread and butter, jam scones and cake.

After dinner Sir Arthur loved to play billiards with me and I always managed to beat him, due to his poor sight. Miss Wiggins would play the piano, a Bechstein and a real beauty, which gave Sir Arthur much pleasure. At other times Miss Wiggins read to me, generally the G. H. Henty books (very popular books for boys before the First World War) while Sir Arthur played his one and only Patience. At 9 o'clock or just after, Sir Arthur would go to bed.

During my early days at Youlbury, Sir Arthur had Duncan Mackenzie staying with him. Mackenzie was an archaeologist and a specialist in pottery and he was a very loyal collaborator with Sir Arthur in Crete for thirty years. He was a giant of a man, over six feet three inches tall but had the softest voice that I can ever remember hearing. Being a Highlander, he would enthral Dennis Haskins, Miss Wiggins's second nephew, and me with exciting ghost stories. He always took a walk round the lake in the late evenings because he maintained that he had talks with the 'wee people' and he saw quite plainly the wraith of Mr Hart who had drowned himself in the lake.

In the earlier days of the excavations at Knossos, Sir Arthur did not have the time to show visitors around the Minoan Palace, so when the famous dancer, Isadora Duncan, arrived one day, Sir Arthur asked Dr Mackenzie to escort her around the site. She was very impressed with what she saw and on arriving at the Grand Stairway of the Palace, she could not contain herself and threw herself into one of her impromptu dances for which she was so well known. Up and down the steps she danced, her dress flowing around her. Dr Mackenzie was very shocked and told Sir Arthur that he did not approve as it was quite out of keeping with her surroundings. Sir Arthur was very amused and from time to time would tease him about this episode.

On another occasion, a party of schoolgirls arrived and before visiting the excavations they were taken to the Villa Ariadne where Sir Arthur and Dr Mackenzie were living. Poor Dr Mackenzie was most put out and complained bitterly as all the girls had peeped through the windows of his bedroom.

Being a Scot, he loved porridge for breakfast, but he had a very thick moustache which acted like a sieve, through which he ate his porridge, much to the amusement of Dennis and myself. I was very sad when, later in his life, he became deranged and died in Italy in 1930.

It was Sir Arthur who introduced me to the beauties of our butterflies and moths which were to be found in abundance among the flowers in the garden and on the large heath patch close to the old oak forest. (This forest was full of primroses and a mass of bluebells in the spring. Sometimes the bluebells looked as if a large piece of blue sky had fallen on to the ground.) Sir Arthur and I would sally forth each day, armed with a butterfly net, a killing bottle and Podger, the name of Sir Arthur's walking stick. Poor Sir Arthur, due to his being myopic, had difficulty in catching the flight of any butterfly, but once it had alighted on a flower, he would then put on his pince-nez and make a scoop with his net – without much success – but he thoroughly enjoyed the occupation as long as I was with him. At night, he would help me to spread the wings of the butterflies and moths. Here, due to his shortsightedness, he surpassed me every time. After tea, he would ask Emma to mix up some beer with sugar and put it in a small bottle and we would go to the oak trees beyond the heath and there brush the beer mixture on to the bark of the trees. After supper, we would go out again with our killing bottle to see what moths had been attracted to our bait. On our way, we would always stop and listen to the nightjars. If we kept quite still, we could watch their extraordinary flight when catching moths. Their peculiar way of sitting on a branch, parallel to it, I have never seen in any other bird, and their song was a lovely purring sound. We never failed to find plenty of moths resting on the bark, suffering from intoxication.

Wherever one walked in garden or woods, there were no straight paths, except outside Sir Arthur's library which gave a narrow view straight towards the part of the Berkshire Downs near to the White Horse and Wayland's Smithy, which seemed to stir deep emotion in his mind. When planning the paths, he marked them out personally with canes for his gardener to follow.

One particular habit Sir Arthur had when out walking with me in the woods and garden was when suddenly some idea or inspiration came into his mind, and then his leisurely pace became almost jet-propelled. He darted ahead, quite forgetting I was with him, and I had to dash after him to catch up. In a second, we would resume our leisurely walking and talking as though nothing had happened. I have watched him from afar, and even in Oxford, he would suddenly dash off at top speed in very short bursts.

Sir Arthur, all his life, had suffered from night blindness caused by his myopic sight, which was very serious, as he would blunder into the trees and bushes unless he carried a lantern or had someone to guide him. On the other hand, in daylight, this was a very great asset in his excavations or looking at minute details of ancient coins, seals, etc. Another strange thing was that he carried his own personal excavation tool; he allowed the nail of his little finger on his right hand to grow a quarter of an inch. This enabled him to remove the dirt from his finds with it. I can see him now; bent over, his eyes close to some small object, working his nail into all the cracks and crannies.

All Sir Arthur's clothes were made for him, and his trousers, regardless of the current fashion, had to be without 'turn-ups'. His boots, (he would not wear shoes) had to be without laces. "Waste of time," he said. So he had a tag at the front and one behind the heel which enabled him to pull his boots on in quick time. He could not bear to waste time, so his shirts were made with stiff turned-down collars attached. Although they came with the necessary slits for a front stud, he merely pushed one end over the other, and his tie was ready-made, so saving time. He was never without his famous stick called Podger. He used it for walking, but it was rather similar in its use to his finger nail; for anything that caught his eye, he would prod it with his stick.

The 'highlight' of the year was his famous children's party, always held at Youlbury on Twelfth Night. What a party that was, for Sir Arthur, so fond of children, really let himself go; he never stopped dancing with all and sundry; it was a sight I shall never forget. The party started at 6 p.m. and finished at 9.30. He engaged an orchestra which played from the wide gallery, but before the first cotillion started, he gave out different coloured ribbons to the boys and girls and when they were all matched, they had to go up to the wide balcony and line up in pairs and then, holding a little girl's hand and at a signal from the orchestra, Sir Arthur would lead us, skipping, round the balcony and down the wide stairs to the large hall. What a sight it was, with all the decorations and the lights blazing in every part of the house. Another thing that caused great delight to the children was when Sir Arthur placed on a table a huge wooden cracker which had a catch in the middle. The object was to choose a partner for the next dance. Six boys would stand at one end of the cracker and six girls at the other end; at a word from Sir Arthur the boys and girls had to insert one hand and try and feel for the hand of the partner of their choice. Having got hold of a hand, thinking and hoping it was the right one, you had to wait until Sir Arthur was quite sure everybody was well clutched, when he would slip the catch and the cracker would split from end to end disclosing all the hands, and there was no changing! You can imagine how disappointed you were when you realized that you were not holding the hand of the partner you wanted.

Half-way through the party came supper, and what a spread there was – sandwiches, cakes, trifles, jellies, fruit, etc. At a given signal from Sir Arthur, Mrs Judd, the cook, would arrive with two large cakes, one for the girls and one for the boys. An African black bean had been inserted in the cakes before baking. This ritual was in order to choose King Minos and his Queen. The boys were lined up in front of one uncut cake and the girls in front of the other. Then Sir Arthur would cut slices for each boy and girl from their respective cakes. They were told that whoever found the bean in their portion would be crowned as King and Queen. It was not long before the result was known and Sir Arthur would lead the boy by the hand to sit on an exact replica of the mahogany throne which had belonged to King Minos at Knossos and the girl was led to another throne. Then Sir Arthur would bow to the King and Queen and the rest of the children did the same. Then followed, with shrieks of delight, the bull game. This took place in the entrance hall, where the floor was in black and white marble blocks designed to look like the labyrinth, and in the centre was depicted the Minotaur – a bull's head with a man's body. Each child had to follow the black marble tiles all over the hall which, in time, brought him or her to the Minotaur, on which Sir Arthur was standing. He would make a bellowing sound like a bull, catch the boy or girl, toss them up in the air and catch them amid shrieks of delight from everybody. The first party I went to must have been in 1913 and I remember I was dressed in a sailor suit. We had one more in 1914 after which the parties ceased because of the war. In one party there was a hitch, for when the cakes were cut and eaten, only the girls' black bean was found – not the boys'! Mrs Judd insisted that the boys' cake had had a bean in it and, at last, a small boy admitted that he had swallowed it.

Sir Arthur decided that a companion should be found for me during the holidays. Miss Wiggins was consulted and came up with the idea that her nephew, Denny Wiggins, should share with me the delights of Youlbury. Denny was about my age or perhaps a year younger. But he did not last long there for he was a bit of a 'know all' and Sir Arthur did not seem to take to him. Then Miss Wiggins thought that Sir Arthur and I would be able to get on with another of her nephews, Dennis Haskins. That, I must say, was a great success, and from then onwards Dennis spent his holidays with Miss Wiggins and me at Youlbury, much to the satisfaction of Sir Arthur.

What lovely holidays we spent in the summer, for Sir Arthur took us all to the seaside; St. Ives, Barmouth, the Isle of Wight and Wales. We visited castles, churches, cathedrals, ancient stone circles, climbing Cader Idris and going up Snowdon by train. During the spring, autumn and winter holidays, Sir Arthur became enthusiastic over stamps and as both Dennis and I had small collections, Sir Arthur never failed to buy, and bring back from London, stamps, mostly unused, for us both, much to our delight. We used to spend hours by the fire, sorting them between us. What a kind and generous man he was, and what happy memories to think back on.

As I was always interested in birds and animals, Sir Arthur bought me some tame rabbits, Belgian hares and Flemish Giants. At first they were kept in a large cage, and then a wire enclosure on one of the small lawns below the house. When I went to Trent College, Sir Arthur used to feed them for me, and in his letters he always referred to them as Bunny 1 and Bunny 2 and sent me little messages from them. One day I was in the Covered Market in Oxford and on going to the Pets' Corner, I saw a young badger. I told Sir Arthur and he at once ordered the car and went to Oxford with a large sack. I think he bought the badger for 30/- and it was put in the old rabbit cage. Every day, Sir Arthur and I would get Mrs Judd to boil up some rice and, to my surprise, the badger ate it all up. After that, I was allowed to go with the chauffeur to Pets' Corner and buy any badger that was there. I think we bought about four, much to the owner's surprise. One day all the badgers escaped but I soon tracked them to a large rabbit burrow close to the veranda which overlooked a small dell surrounded by thick tall fir trees. The test came when we put a bowl of cooked rice at the most likely hole. Sure enough, next morning it had all gone, and we saw signs of fresh bracken having been pulled down the hole, as badgers like to keep their sets clean. Next evening, after supper, we sat watching the main hole and the bowl of rice. Sure enough, the two young badgers came out and squabbled over the rice. As soon as I saw them, I nudged Sir Arthur. He at once put on his spectacles and was able to see them well. We watched them for many nights but, in the end, they moved away.

As I have already said, Sir Arthur could, at times, become very angry. Not far from the house, beside the large pine trees, were the 'upper lawns', on which were a tennis court and croquet lawn and pavilion. Sir Arthur always liked a game of croquet, but he was very handicapped by his poor sight and he had to use his pince-nez. This particular afternoon, I was not in the mood for croquet; I wanted to ride my bicycle with Dennis round the paths in the woods. I knew that I had to play, as Miss Wiggins had chosen Dennis as her partner, leaving me to play with Sir Arthur. I was in a mood best described as 'agin the Establishment'. I started by missing easy shots, and I could see Miss Wiggins was not impressed. At one stage of the game, Sir Arthur had positioned his ball just right to go through the hoop the next time, provided I knocked Miss Wiggins's ball away. It was an easy shot for me to do but as I was feeling very sulky I took a blind shot and missed Miss Wiggins's ball and knocked Sir Arthur's ball across the lawn. It was just too much for him; he turned on me, his eyes blazing.

"You bad-tempered, ungrateful child. Go home, go home, at once. I don't want anything to do with you." I took to my heels and fled down to the house and up to my bedroom where I rang the bell for Ada.

When she came in and saw that I was in a flood of tears, she exclaimed, "Oh, Master Jimmie, whatever is the matter?" I told her what had happened and that she must pack my trunk and I must go back to Blagrove, never to return. She replied, "Don't be silly, Master Jimmie, Sir Arthur does not mean that at all. He means Youlbury home, not Blagrove." After some more tears, and Ada comforting me, I went down to tea in the drawing-room, and, to my surprise, Sir Arthur behaved as though nothing had taken place between us. I can't say I enjoyed my tea and cake, and when Sir Arthur got up to go to the library, I followed him and told him I was very sorry that I had behaved so badly. He promptly kissed me. I wonder what his fellow archaeologists would have said about that side of his character.

* * *

I must now pass on to the second stage of my education which was at Trent College, a Church of England School. The Headmaster was the Revd J. S. Tucker. Sir Arthur took me up to Long Eaton and had a long talk with him and then said goodbye to me outside the school gates. This was the September term, 1916.

I would like to get through this part of my life as quickly as I can. Trent College was a good school with good discipline. I started in the second form, but it was soon realized how unacademic I was. I soon gave up Latin and French and I was poor in Algebra and Geometry which resulted in my being moved, after one year, to the 'Remove', then on to 'Shell', and finally into the 'Army' (or Barmy) Class. Owing to my mastoid, I was barred from Rugby, boxing, etc., but I played a little cricket. The result of all this was that I became a 'loner'. I was sixteen years old when the war finished, so all my training in the Army Class was wasted. The Headmaster and Sir Arthur agreed that no useful purpose would be served by my staying on at school, and I finally left in the spring term of 1919.

Throughout the war, during the holidays at Youlbury, I met many interesting and famous people. A good many of these people lived on Boars Hill, as Oxford was only three miles away. Dr Bridges and John Masefield were the major poets, living not a mile apart, but there were also minor poets living scattered about the Hill. Then there were Professor Gilbert Murray, the founder of the League of Nations, Lord Berkeley, Lillian Macarthy, the actress, Sir William Boyd Dawkins, the anthropologist and many others who were frequent or occasional visitors, as was Mortimer Wheeler.

Sir Arthur told me this story when Sir William Boyd Dawkins came to Youlbury after exploring the caves in Patagonia in the Argentine. It seems that one day when he was in a very dry cave, he found the remains of a brontosaurus from which he pocketed a small piece of its hide. This small piece he kept in his waistcoat pocket and delighted his friends by showing it to them.

Dr Bridges, the Poet Laureate, I saw many times for he often came to tea at Youlbury, or I would go with Sir Arthur to have tea with him and his family. On one occasion, Sir Arthur, Dr Mackenzie and myself were invited to tea at the Bridges. Promptly at 6 o'clock Sir Arthur rose from his chair to return to Youlbury for his customary letter-writing and I stood up too, but it happened that Mackenzie was engrossed in conversation with Mrs Bridges and he wanted to stay on, and as we moved towards the door, leaving Mackenzie sitting on the sofa, Dr Bridges, seeing the situation, called to Sir Arthur, "Evans, do you mind calling your friend off."

There is another amusing story concerning Dr Bridges during the time he lived on Boars Hill. He had a strong premonition that his house would be burnt down. As he could not drive a car, he would order the local taxi to take him to Oxford when he needed to go there, and on the journey home he would ask the taxi driver to pull into a lay-by half-way up Hinksey Hill. From here, looking towards Hen Wood, he had a clear view of his house in the distance. He would lower the window, look out towards his house and then tell the driver to drive on. This went on for some years, but one fine morning, returning from Oxford, the taxi driver pulled into the lay-by, Dr Bridges lowered the window and popped his head out as usual, and what do you think he saw? His house in a blaze of fire! Dr Bridges's only remark was, "Ah, at last! Drive on!"

On another occasion, while having tea with Sir Arthur and myself, on the terrace at Youlbury, Dr Bridges was discussing with Sir Arthur the books and poetry that Masefield had written when suddenly he came out with, "Not a bad fellow, but he will bring his wife with him when he visits." I nearly died with laughter. Mrs Masefield was rather eccentric in her dress; she wore an old hat and a long coat and was often seen alighting from the bus carrying her groceries in two large fish baskets.

There was no getting away from it – Dr Bridges was a strange man. You would never know, when you met him in the woods, whether he would stop and talk with you, or walk past without saying a word. One day I was with my dog, Jock, shooting rabbits. I had shot four and, as my old dog was feeling tired after all the hunting, I decided to sit down and have a rest. After a few minutes I saw Dr Bridges, wearing a Sherlock Holmes type overcoat and a large-brimmed hat, coming along the path towards me. Seeing me, he stopped and asked if I had had good sport. I told him I had, whereupon he stooped down to look at the dead rabbits. This was too much for Jock, who grabbed his trousers and gave them a shake. You should have heard the language – not at all poetic! I had never known Jock do anything like that before, but I expect he thought Dr Bridges was going off with the rabbits.

When he became Poet Laureate, a great number of people were drawn to his house and looked into his garden, so he instructed his gardener to place a notice on his garden gate. It read;

NOTICE –
NO VISITORS IS ALLOWED INSIDE OF THIS GATE.

and somebody had signed it for Mr Bridges – 'R. Bridges Pots larret'.

Lord Caernarvon was another visitor at Youlbury and stayed several times while I was there. He was a keen motorist, but unfortunately, he had had a very bad smash outside Paris which resulted in his face being badly damaged. The doctors rebuilt his face, but it left him scarred, as though from smallpox. When he bought his first car, long before the 1914-18 war, it was his great aim to drive up Beacon Hill, an ancient hill fort on his estate at Highclere, near Newbury. He had tried several times, to reach the top, but could never make it. One day, Sir Arthur had motored over to visit him and Lord Caernarvon suggested that they should have another try to climb the grassy slope which was nearly one thousand feet high. This time, he changed tactics and went up backwards and, to their delight, reached the top; they had overcome gravity!

Meanwhile, the Youlbury Scouts had outgrown the old HQ so the upper lawns were handed over to them, but within a very short time even this HQ became too small. Sir Arthur then decided to build a purpose-built one on the very ground that I had first met him all those years ago. Sir Robert and Lady Baden Powell came to lunch at Youlbury and then on to open the new HQ. The opening was a grand affair. There was a large gathering of people and several other troops were drawn up on the parade ground to watch the Youlbury 29th North Berkshire Troop's flag raised. There was a constant demand from boys wishing to join us, but many of them lived in neighbouring villages, too far away for them to walk. But here again Sir Arthur's generosity was seen, for he even bought them bicycles so that they could come twice a week to do scouting. He also gave us the use of a room known as the Solar

Room at the top of the house, above the servants' quarters. It was a large room which contained a piano and which, in the old days, was used as a playroom by any children who were staying at Youlbury. Sir Arthur was keen that the boys could use this room on winter nights, where they could sing all the Scouts' songs with the aid of buns and cocoa! Frank Gillams's fiancée, Winnie, from Oxford, played the piano.

At last, after dragging on for four years, the war was over on the 11th November, 1918. Sir Arthur, at once, decided to build a war memorial to all the men and boys he had known at Youlbury and who had been killed. He decided that he would make a new path to the memorial which he called the Peace Path. Sir Arthur's house was built on a small escarpment, overlooking the lake, directly facing south. Fir trees sheltered the east, north and west from the winds. On the terrace towards the east was a group of tall, closely-growing fir trees which dipped down from the sunlight on the terrace into a dell with only a little sunlight filtering through. This new path went from the terrace down into the dell and, in due course, climbed up and broke out of the darkness of the trees again into full sunlight. This dark part of the path represented the dark and terrible years from 1914-1918. Two scarlet oaks, symbolising peace again, were planted beside the winding path which led to the beautiful memorial which faced south, looking away over to the Berkshire Downs. Behind this memorial, were planted cypress trees, and inscribed on the memorial were the following words:

Horas
Non numero niso serenas
– Sundial

I count the sunny hours alone
1914 - 1918
In loving memory of a Youthful Band
Who played as Children
Amid these woods and heaths
And shared at Youlbury in joyous hours
In the Great War
For their Country and for Mankind
They fell before their time
But wherever they now lie
Here they are never far

On the left-hand side in gold

Not for this Motherland alone
Their youth, their love, their life they spent
The glorious halo round their brows
Shines on a wider fundament.

On the right-hand side in gold

Theirs was a loftier sacrifice,
For after years and all mankind
Faith and humanity should bind:
They fought that wars themselves should cease
And they have entered into peace

My brother, Gilbert, had joined up in Canada in 1914. He was wounded and Mentioned in Despatches twice. My brother Dick, four years my senior, was very badly wounded in both legs, but he survived and I remember visiting him in hospital in Leicester while I was at Trent College.

I was in the OTC at school and I presented to my mother a photograph of myself in uniform, thinking how proud she would be to see how keen I was to serve my country, but I was very taken aback because when she saw it, she burst into tears. I now know why.

My brother, Gilbert, having been demobbed in Canada, returned to Blagrove in 1920, undecided what to do. My Uncle Will, who lived on Boars Hill, and, as I have already mentioned, had been for many years engaged in importing pedigree bulls and sheep to the Argentine, asked Gilbert if he would like to re-visit the land of his birth. If so, he could take out some sheep and bulls. Gilbert jumped at the idea as it meant a first class return trip on the *Highland Laddie* and all he had to do was feed, water and generally look after the animals.

This, of course, necessitated his having a visa to enter Argentina, as he had a Canadian passport. I decided to go with him to the Argentine Legation in London. We were ushered into the office and there, seated behind a large table, was a very short gentleman. Gilbert told him what he wanted and that he would only be three weeks in the Argentine before returning on the same boat in which he had gone out. The Consul then asked for his name, age, birthplace, etc. When Gilbert told him he had been born in Buenos Aires, the fun started.

"You are not English, or Canadian, but an Argentine citizen and, what is more, you have not done your military service in my country and therefore you are a deserter." That did it; Gilbert just 'blew his top'. His summing-up of all men was that they were either 'a lovely fellow' or 'A God-damned son of a bitch'. And the latter was just what he called the Consul. I became alarmed as they were both on their feet ready to 'have a go'. So I got them to sit down and talk things over. The Consul said he would grant him a visa but that he would be arrested and have to do eighteen months' service in the Argentine army. I told the Consul about Gilbert's war record and this certainly impressed him and he relented and gave a guarantee that he would be free to enter the country, on condition that he returned to the UK by the *Highland Laddie*, but if he stayed one day longer he would be arrested and charged as an Argentine deserter. Anyone born in the Argentine will always remain an Argentine citizen.

During the 1914-18 war, Sir Arthur became involved in a furious argument with Sir Alfred Mond, who wanted to take over the British Museum as the headquarters for the Air Board to house civil staff for war purposes. This idea infuriated Sir Arthur and he started a hue and cry in the Press, etc. Due to his arguments and persuasion, it was stopped. Sir Arthur's remark to me was that he could not bear the idea of Sir Alfred 'prancing through the halls of art'.

In 1916, Sir Arthur became the President of the British Association, one of the highest honours that could be bestowed; furthermore, he remained in that position until 1919, which was a record. While on holiday at Barmouth with Sir Arthur, Miss Wiggins, Jack Evans, Sir Arthur's nephew, and Dennis Haskins, we met a man who, when he found out who Sir Arthur was, took me aside and said, "I would rather be President of the British Association than the King of England."

That was a wonderful and interesting holiday. We hired a car and visited all the beauty spots in North Wales; we climbed Cader Idris, leaving Miss Wiggins in an hotel on Lake Tall-y-llyn. On reaching the top, in glorious weather, Sir Arthur refreshed us all with ginger pop.

It was during these holidays in Wales that Sir Arthur arranged with the gardener that the peaches and nectarines, grown at Youlbury, were to be put in a large wicker basket with softly-lined pockets, and closed with a rod and lock, and despatched by train from Oxford to Wales, where they would be collected by car and enjoyed by all of us.

It was in 1918 that I was confirmed in Christ Church Cathedral in Oxford. I shall never forget that day. It was 21st March, when the Germans burst through our lines on the Western Front and got very close to Paris. It was a near thing for everybody and even the cooks and communication troops and reserves were thrown into the large gap caused by the Germans. Thank goodness, we held in the end. I was among a whole group of children being confirmed. When the ceremony had been completed, the Bishop of Oxford told us what had happened and asked the whole congregation to kneel while he offered up a special prayer to save us from this catastrophe. That memory of Christ Church will never leave me.

Soon after this, there followed a very interesting period in my life. It had become clear that the Germans were desperate for manpower for their armies, so they conscripted all boys from fifteen years upwards into their armies, from Serbia and Montenegro. These people got wind of this in time, so they collected up everybody of near military age and marched them over the Carpathian mountains in winter, down to Triest. From there, the navies of Italy, France and Britain took them to France and then to England and, in due course, many of these boys arrived at Youlbury where Sir Arthur housed them in tents, etc. You can imagine their joy at being addressed in their own language by Sir Arthur. Whilst at Youlbury, they carved lovely wooden toys of their countries and presented them to Sir Arthur, who hung them in the

South Hall for many years.

It is very difficult to remember the sequence of events that happened fifty years ago among the Slavs, Slovenes, Croats, etc. but these people started to rebel against the Germans and Austrians, which was encouraged by the Allies with the supply of weapons, etc. The Germans had to withdraw divisions to contain these outbreaks.

It was, I think, in July 1917 that, one day, Sir Arthur told me that a delegation from the War Office would arrive at Youlbury and that he would not be able to play billiards with me after supper. I was not told who these men were and even Emma did not know. She only knew that they were coming at 11 p.m. and that she was to take refreshments into the library. I, being curious, decided to stay up and stationed myself on the balcony overlooking the hall with the Minotaur for company. At 11 p.m. the bell went and Emma opened the front door and led a group of six men, dressed in dark suits and overcoats, into the library. I asked Emma what language they were speaking when she had gone into the library with the refreshments. She told me that Sir Arthur was talking 'foreign', but she heard English as well – all very mysterious cloak and dagger stuff it seemed to me. They left by car in the early morning. At that time I liked to get up early so Ada arranged that my plate of fruit was left by my bed the night before. The next morning was a beautiful one and I went on my usual walk down to the lake. It was about 8 o'clock. As I walked round the south-east side, I noticed that there was an elderly gentleman sitting on the iron bench in the sun. I must say I was surprised to find a complete stranger there at that time of the morning. As I approached he said, "Good morning. Are you Jimmie Candy?" He must have noticed my surprise when I answered "Yes". "I am Tomas Masaryk and I am staying with you and Sir Arthur." Then I knew that he was one of the party that I had seen crossing the hall the previous night. Masaryk asked me to join him on the seat and there we talked until it was time to return to the house for breakfast. He was quite a small man of about sixty, with a goatee beard and a turned-down moustache and he wore spectacles. He made a great impression on me and I was at once drawn to him and felt quite at ease with him. He stayed with us for a whole week, spending most mornings in Oxford or in London, and the evenings in the library, talking to Sir Arthur. He suggested that I should meet him at 7 a.m. each morning, weather permitting, on the same seat down by the lake. He told me his life history and why he was in England, and the more he talked the more I was drawn to him. It seems he was the son of a coachman, self-taught, spoke English and French well, and had a burning ambition to free his fellow countrymen from being ground down by the Austrians. He was thrown into gaol by the Austrians. His sister had been raped in front of him by the gaolers and all sorts of terrible things were done to him. Eventually he was released. Then came the Great War, and, during that time, he worked against the Germans. The Allies decided to smuggle him out, first to France and then to England, as they could see that he would be the very person to lead the Slav people to freedom. Sir Arthur was well known to the Slavs; he spoke their language and supported them against the Austrians, when he was a young man. Also, he had travelled throughout these regions before the war, so much so that the Austrians had thrown him into prison for supporting them.

One morning, while we were sitting on the seat, Professor Masaryk told me his plans for the future, when the war was over. He would form a new partition, made up of the various races and the new country would be Czechoslovakia and he would be their first President! What is more, when that was accomplished, he would send me some stamps and, sure enough, he did. The last morning we were together, he told me he was very interested in the Scout movement and the whole hour was taken up talking about Scouts; what they did, how they were formed and, particularly, from which classes of the population they came. When I told him that the movement covered all classes, and that it was a great opportunity to mix all boys, rich or poor, under the Scout flag, Masaryk made a very significant remark.

"I cannot hope to have a real democratic country, without the help of the Scout movement." How true that is. He left the next day and I did not see him again. From time to time, Scouts from Czechoslovakia came to the Youlbury Camp.

There was an incident that showed what the Czechoslovakian people thought about President Tomas Masaryk. Here in Abingdon, a young woman applied to me for employment in my office, as a secretary. This woman was Czech, of Jewish descent, and had fled her country at the outbreak of the 1939-45 war. One day I told her what had passed between Masaryk and myself. On hearing the story, she became very excited and asked me if she could visit the spot where these conversations had taken place and could she bring another Czech woman with her. So one evening we set off to Youlbury which, by then, was empty, as Sir Arthur had died in 1941. I took them to the very spot – yes, even the iron seat was still in the same position. It had the most moving effect on these two women, for they burst into tears as they stroked the old iron seat. Even I had to turn away with tears in my eyes. I knew how the Czechoslovakian people worshipped Tomas Masaryk, for after his death he was buried in a forest and every year the people came in their thousands and placed flowers on his grave. During the war, the Germans tried their utmost to stop them for fear of stirring up emotional feelings against the Germans, but unsuccessfully, so they then decided to seal off the grave by erecting a high electric wire fence, but the people still came and they brought catapults with them to shoot the flowers over the fence on to his grave. When the Communists seized power in Czechoslovakia they destroyed every monument to Tomas Masaryk and only left one at Hodonin in Moravia where he was born. This they tried to pull down but the inhabitants resisted all attempts.

* * *

Now came the time that I had to move away from Sir Arthur and Youlbury, to start my new life. When I left Trent College, my ear was still discharging and the doctor thought it advisable for me to have an open-air life. Sir Arthur thought that I should learn farming on a large farm up on the Berkshire Downs in the village of Upton, where I spent two years, 1920-22, living with a Mr and Mrs Humphrey, for which privilege Sir Arthur paid them £200 a year. It was my first enjoyment of complete freedom and for the next two years I had a whale of a time. I never did any work, but just watched the farm labourers doing their work. I had always been very fond of shooting and this farm consisting of over 1,000 acres, had plenty of rabbits, partridges and pheasants. I spent the winter months shooting, the summer months going to tennis parties and dances, and at the weekends I returned to my family at Blagrove and rejoined my brothers and sisters, or I spent the weekend with Sir Arthur.

On this farm there were four other men, all paying their £200 a year. I was the youngest and the other three were all ex-service men who thought they would take up farming as a career. One of them, who had been badly wounded in the war, was Frank Madge, whose grandfather, Sir William Madge, was the founder of the newspaper *The People*. He and I became very great friends indeed, and this friendship continued until he died, in 1962, and still continues through our two families. What a happy period that was; not a cloud in the sky. My father thought it was a waste of time and not the way to learn farming. But, as I said, I was enjoying perfect freedom for the first time in my life – not bad for a seventeen-year-old.

During the last war, Frank Madge and his wife, Woggie, and their four daughters lived quite close to us at Shillingford. Before and after the war they lived at East Grinstead and when Kitty, my wife, and I visited them there, it was a pure joy. What a happy, jolly family they were; due, I think, in part, to such good communication with each other, which is the basis of a happy family.

After consulting another ear specialist in London, it was thought that I ought to live in a warm dry climate, and Sir Arthur suggested the Argentine, and he went to see my Uncle Will. He agreed that I could go out with him to Buenos Aires with a cargo of pedigree Lincoln rams; also he would do his best to find me employment on an English estancia, so that I could learn the language.

Now came the time for me to step out into another world. On 22nd January 1922, I set sail from Liverpool with my uncle on the *Haliartus*, an old cargo boat, bound for B.A. with a cargo of railway engines for the Great Southern Railway of Argentina. My uncle had twenty pedigree Lincoln sheep, which I had to help look after. These sheep were installed in wooden sheds amidships. They were well protected against sea water as they had long fleeces which were vital for their sale in B.A.

Sir Arthur saw me off at Youlbury and dear Ada kissed me and cried at the same time. I then went to Blagrove to say goodbye to my family. Poor mother was very upset at my going. She said that out of her six children, five had gone abroad; Gilbert to Canada, Ethel to Columbia, Madge to Latvia, Dick to Canada and myself to the Argentine; only Martin was left to comfort her. As a parting gift mother gave me four five-shilling pieces, dated 1888, which had been given to her as a wedding present and she had brought them all back to England in 1900. Mother told me that these five-shilling pieces would being me back to her again, safe and sound. When I finally left Argentina and came home in 1933, I was able to show her only one of the pieces, but more of that later.

We had not been at sea long, before we struck a violent storm. My first experience of being on a ship and being sea-sick, I shall never forget. I really thought the ship would never survive. The worst part was when we were in the Bay, for there, we had to hove to and put the ship's nose into the storm and ride it out. My uncle and I had a small cabin as we carried no passengers. My uncle just wedged himself in his bunk and read a book – no storm disturbed him. In fact he had crossed the Equator eighty times – mostly to the Argentine with pedigree stock – by the time he died aged ninety-six. Meanwhile I was as sick as a dog and just lay in my bunk, watching my trunk dashing about the cabin floor as we rolled from side to side. Up forward, we had securely tied a number of empty wood wine casks which were bound for Vigo. Every time the old ship put her bows into the big waves she seemed as if she would never come up again; there would be a lull, a tremble all over, then up she would come with a crashing noise as the water rushed along her decks and back into the sea. It was not long before this sort of treatment to the empty wine casks caused them to break up and be swept among the winches and over the side. The noise was terrific and I thought that the ship would never survive, but, looking at amy uncle, wedged in the corner of his bunk reading a book, reassured me and made me realize that things were not too bad.

Our cabin was alongside 'Sparks' and when we heard him call the Captain that he was receiving an SOS from a sinking ship, that really put the wind up me. The Captain asked 'Sparks' how far off the ship was and was told two hundred miles from our midday point, and that there were two ships closer to the stricken ship which were going to her aid. Afterwards, we learnt she went down with some loss of lives. Within an hour another SOS came through from a ship on the rocks at Brest. All that day, I watched our two trunks dashing about the cabin floor, first under my bunk and then under my uncle's. Why did I ever come to sea? I was sick for two days. I remembered so well the song of Kipling that I used to sing with my mother when I was a small child.

> When the cabin port-holes are dark and green,
> Because of the seas outside;
> When the ship goes wop (with a wiggle between),

And the Steward falls into the soup tureen,
And the trunks begin to slide;
When Nursey lies on the floor in a heap,
And Mummy tells you to let her sleep,
And you aren't waked or washed or dressed,
Why, then you will know (if you haven't guessed)
You're 'Fifty North and Forty West!'
Why then you will know (if you haven't guessed)
You're 'Fifty North and Forty West!' *Rudyard Kipling*

Next day, the seas abated, but there was still a great swell. Nevertheless, my uncle insisted upon my getting up and dressed and having some breakfast so that I could help him feed the sheep. At the time, all I wanted to do was stay in my bunk, but my uncle was right. Making the effort was the only way to get over the sea-sickness and get my sea legs. Sometimes I fed the sheep with hay and crushed oats; other times I fed the fishes! The poor sheep ate nothing during the height of the storm, but when it abated, they were soon back to normal – fresh bedding and water did wonders for them. At other times, I have crossed the Bay when it was as smooth as a mill pond. When we arrived at Vigo, we unloaded the remains of the wine casks. From Vigo to B.A. we had lovely weather, taking thirty days to get there.

Uncle and I had all our meals with the Captain and his officers. As we did not carry any refrigeration, we had only salt beef, etc., to eat. Everything was salty, except the tinned fruit, jam and cornflakes. We had an awful job to stop the crew from bringing the left-overs from their plates and throwing them into the sheep troughs. Sometimes we even found bones there. They were under the impression that the sheep would like a change of diet from crushed oats, mangels, and hay.

The Captain kept the ship well away from the usual shipping lanes. In fact, we never saw any ships for days on end. Eventually, when we were some distance from the Brazilian coast, we spotted land. It was the top of a mountain sticking out above the sea, called St. Paul's Rock; not more than an acre of barren rock above the sea and not very high. I thought what a God-forsaken place it looked; no water there and uninhabited. What we did see round the ship were plenty of sharks, also flying fish which came on to our decks at night, attracted by our cabin lights. The ship's cat had a wonderful time!

Two days later, we could see, on the horizon, a large island off the Brazilian coast. This turned out to be Fernandod Noronha, a large penal settlement, and what a forbidding place it looked. No escaping from there, for the sea teemed with sharks. On coming down the Brazilian coast, although out of sight of land, I did notice that the sheep sensed that land was not far off. They became restless and seemed to sniff the air. Also, birds I had not seen before, came aboard and rested on the ship for, perhaps, half a day before taking off again. Beautiful butterflies also settled on our ship as we neared the Uruguayan coast. Many large liners from all parts of the world were making for the River Plate, on their way up to B.A. or leaving for Europe. We spent one day in Montevideo, the capital of Uruguay. It was lovely to feel the ground beneath our feet again and stretch our legs. How nice it was to have a really good meal with plenty of fresh vegetables and fruit. We were there for a whole day, looking round the capital and I heard Spanish for the first time.

That evening, we sailed up the River Plate and, in the morning, docked in B.A. We moored in a new dock with all modern equipment. The gangway was put down and I walked off the boat on to Argentinian soil on February 16th 1922. I could not help remembering my parents' description of how they came ashore in 1888; first by rowing boat to a large cart drawn by a team of oxen, which were standing knee deep in the water and thence to the shore where they were deposited with their belongings.

I watched with great interest how they would unload the twelve rams. Very simple. The quarantine authorities led a most decrepid-looking, flea-bitten ewe up the gangway to our sheep pens. We then knocked out the sides of the pens and let the rams free. They had one smell of the ewe, which promptly turned round, and the whole lot set off following her across the deck, down the gangway, in single file, right into the quarantine pen and they never saw the ewe again – very hard luck after such a long sea voyage.

* * *

So there was Buenos Aires, the country where I was going to make my fortune! It was certainly a fine city – the Paris of the South; wide streets, large hotels and excellent restaurants, fine parks and gardens and Banks of every nation. What struck me was how well the Argentine men and women dressed. The best place to see this was in Calle Florida, which was closed to traffic between five o'clock and six o'clock each day and it was here that the people paraded, talked, flirted and showed off their finery.

Everywhere you could hear English spoken, for at that time all the railways, telephones, power stations, port facilities, shipping, etc., were under British or American control. There was a British population of 60,000 in the country. The city had large department stores like Harrods, Maples, Mappin & Webb and many others, where all the top staff were British or European and the lower ranks were Argentine. There were two English daily papers as well as excellently run Argentine papers like *The Nation* and *Prensa*. At the time of my arrival, things were changing; no longer did you see English or Scottish engine drivers and more and more of the Argentines were taking the top jobs from the Europeans. There were never any problems of racial trouble between the different races in the country. Every year, boat loads of people, mainly from Spain and Italy, found work in the towns or went into the interior to work on the farms where there were vast herds of cattle, and cereals were grown. The interior was known by the Argentinians as 'El Campo', but among the English-speaking, the word became 'Camp'.

The time had now come for me to find employment as my uncle would soon be leaving for England after the sale of the rams. Before leaving England, I had planned that when I arrived in the Argentine, I would look for a job on one of the large estancias. There were many English-owned estancias like Bovril and Leibigs, etc. which had many large farms dotted about the Argentine, with English managers. At the interviews I had, the managers of these estates were quite sure that they could give me employment in a junior position. They even told me my salary and conditions of employment. Then came the vital question – "I take it that you can speak Spanish?" I had to admit that I could not. "Mr Candy, I am afraid I cannot offer you this post, as you could not give instructions to the peons." This became the familiar pattern of all the interviews with other companies. I was becoming desperate. I could have found employment in the British or American banks or with the British-owned railways, but I was determined to keep to my object of working on the estancias. I was told by a friend to call on a certain Miles Passman, a German, who owned an estancia on the Western Railway line. This time I changed my tactics by admitting to him straight away that I could not speak Spanish, but that I was determined to learn, if he would give me a chance, even if I had to live and sleep with the peons. To my delight, he agreed to my request and said he would tell his manager of my arrival.

* * *

On arriving at the railway station, I was met by the manager, or major-domo, Alex St. Clair Simpson, who greeted me by saying, "Why the bloody hell did Mr Passman send me a person who could not speak a word of Spanish?" Not exactly a delightful way of making me feel that I was wanted! I was driven by car to the estancia La Isabel, which was to become my home of many months. I had taken the precaution of buying a camp bed in case there were no beds there. After lunching with Mr and Mrs Simpson and their two small children, I was handed over to the Swiss under-manager, or secundo major-domo, whose name was Staffer, and he led me to the peons' quarters. The building consisted of mud walls and floor and covered by a tin roof. There was one large room with small windows which was used by the peons as a bedroom and here about a dozen men slept. There was another room which was used as a dining-room. When more buildings were needed for houses, etc., the only things required for building were earth, straw and a small number of horses confined to a small fenced-in yard, and plenty of water. All that was needed was a small boy on a horse who would proceed to keep the horses moving round and round in the enclosed space for at least a day, churning up the mud and straw. When this reached a certain consistency, the gate was opened and then men would move in and scoop up the mud which they put into wooden moulds in the shape of bricks and leave them in the sun to dry. They were then placed in a kiln and fired. This was done by stacking the bricks and covering them with mud. The wood fire was then lit and so the bricks were fired, and these bricks would last for some years. As the earth was purely loam, it was completely free of stones and I never saw even a little pebble all the time I worked in the camp.

Now back to the sleeping quarters on my first night at La Isabel. Soon after dark, the men went to their dormitory. Some had camp beds similar to my own, others just slept on the rugs from their native saddles and covered themselves with their ponchos. First they removed their top boots, or 'alpargatas', a kind of rope shoe, mainly used in summer. Then followed their 'bombachas', a loose kind of trousers rather like the Turks wear, only they had three small button fastenings just above the ankle. These bombachas were very comfortable to wear, especially when on horseback. Then followed their short Eton-type jackets, which they used as a pillow. But most important, their large knives in their scabbards were placed nice and handy. Then came my turn to go to bed; being a well-brought-up Englishman, I washed my face and hands and teeth and changed into my pyjamas. All this certainly caused great interest among my fellow sleeping companions. They crowded round my bed jabbering away in Spanish and, of course, I did not understand a word. It was some months afterwards, when I had learnt quite a lot of Spanish, that I found out that they thought either I was dressing up to have dinner with the major-domo or that I was off to the village for a dance. I shall never forget that first night with twelve men all snoring their heads off, while I lay in bed wondering what Sir Arthur would say, or how shocked my Ada would be for her Master Jimmy!

The next morning soon came round, but there was no peach, apple or grapes for me! I was awakened by a noisy bell just half an hour before the sun rose over the horizon of a perfectly flat land. All work started at the first peep of the sun and ended just as it dipped below the horizon in the west.

After dressing, I was handed the traditional 'maté'. This consists of a gourd, hollowed out and dried for several weeks. Into this gourd is placed the bombilla, a hollow silver or nickel tube, at the end of which is fitted a small strainer, the size of a thumb nail, perforated with many small holes. The gourd is filled with the famous maté, a type of green Paraguayan tea (yerba). To this is added hot water from a kettle and then it is handed round. On sucking at the bombilla, you draw up the maté, which has a strong bitter taste. When you have had your drink, you hand the bombilla back to the server who at once fills it again from the kettle and then hands it to another man and so on until they have all been satisfied. From time to time, more of the yerba is added to keep up the strength. This drinking of maté is carried on whenever there is a supply of hot water. In the evenings, the men sit round an open fire and drink maté, while talking, singing or telling of past exploits to one another. There is no doubt that maté has very sustaining powers besides being of medicinal value. Some people like to drink it with a little sugar.

This was how my life as a peon started. All the peons, whether on horseback or on foot, carried knives. One for everyday purposes and the other for Sundays, when they liked to visit the villages. The former had a blade of about eight inches in length and was very sharp. This was kept in a scabbard and stuck in the belt behind the back, with the handle to the right (or left for a left-handed man). The latter was a more expensive knife, generally with a silver handle and a more decorative scabbard. Furthermore, there was a guard on the knife, similar to a sword, which would prevent any danger to their thumb or fingers, if they were involved in a fight. The everyday, all-purpose knife was used primarily for cutting up meat at meal times, skinning dead animals, cutting leather when making halters and reins, or any other work that needed a sharp knife. The peons wore no braces, but bound round and over their bombachas trousers, a long cummerband and on top of that was worn a large wide leather belt into which the knife was stuck.

Most of the peons wore shirts but without a collar and no ties. They wore a black scarf tied round their necks which in hot windy weather could be used to cover their mouths and noses to keep out the dust.

Meat was the main diet – morning, noon and night, boiled or roasted. We used to kill a cow every day in order to feed all the staff, peons and their families. The meat was boiled or, as we preferred, cooked over an open fire. A large joint was put on a grid or harrow over the fire. When the cook gave the word that it was ready, we would go up to the meat, draw out our knives, choose the piece we wanted and cut it off, hold it in the left hand in which we already had a 'galletta', or hard biscuit, at least six months old, which stopped our hand from being burnt. We put one end of the piece of meat into our mouths and cut off the rest with the knife close to our faces, taking great care not to cut off our noses.

After a few days sleeping with the peons, I approached the secundo major-domo and asked if I could be moved into better accommodation as there was available a small two-roomed bungalow about a hundred yards from the peons' quarters. The following day Staffer told me that I could move but I would have to share the bungalow with the Capatas, who was in charge of the pedigree stock, which were housed in the galpon (a very large cow stall). What a relief this was, as the bungalow was well made, with a tin roof, and the walls were of mud bricks, whitewashed, and the floor was concrete. At the back was a small washplace and a clothes line. What luxury! For the next six months I worked in the galpon with a Spaniard and the Capatas, who was called Bonefacio. He was of Indian extraction, small and bad-tempered. On the other hand, he was considered a very good Capatas. His knowledge of pedigree stock was above average, and he was most experienced in the preparation of the bulls and heifers for the Palermo Show in Buenos

Aires which was similar to our Royal Show in England.

He certainly took advantage of me and I was ordered to do many menial jobs for him, such as cleaning his boots and serving him with maté whenever he wanted it. He had no words of English nor I of Spanish, but he certainly made sure that I would learn. Many a time he would fly into a rage if I brought him a bucket when he had asked me to bring a fork!

The galpon, where the pedigree heifers and bulls were kept, had to be kept clean at all times. Our day started with the Capatas shouting to me to get up at 3.30 a.m., serve him with maté and be in the galpon by 4 a.m. All the bulls were chained but there were four loose boxes containing the young heifers. The animals were roused and all the dung removed and fresh bedding brought in. We then watered the cattle and fed them a meal of crushed oats, cake and bran. While they ate, we had our breakfast in the peons' quarters, but we were back at work again within half an hour. The bulls had halters on and a rope was passed through the ring in their noses to enable us to have complete control of their movements. These bulls ranged from two to three years old and were going to be shown at the Show in the spring and then sold.

We had a small corral into which the bulls were tied two at a time. Then, armed with hoses and buckets, we washed them from top to tail with a soapy solution. Then came a very important grooming with brushes to all parts of their bodies and particularly their heads and tails and to cover up any defects from the eyes of the judges. In time, I became good at hiding any faults, by fluffing up the hair and brushing it certain ways. When all the bulls were washed, we set off, three at a time, for exercise and training. We would walk them a quarter of a mile, making sure that we walked close to the bulls on the left side of their heads and at the same time keeping their heads parallel to their backs. No bull would win a prize with his head down, which is the natural way that a bull walks. How my arm used to ache and I longed to lower it to stop the pain, but Bonefacio would spot it at once and yell my name. In time, of course, I got the knack, and the bull learnt too, that he had to carry his head in a certain position. Then came another important lesson that the bull had to be taught. This was to stand still for as long as you thought necessary, in a certain position, i.e. the forelegs close together and the right back leg slightly forward. These two positions showed off the bull at its best in front of the judges and for photographs to be taken. We carried a small stick just for tapping their hoofs to get the proper position. Very soon the bulls knew the rules and usually we had no trouble. Furthermore, we had to walk them at a set pace and they seemed to know that they were a race apart from all other bulls.

All this fell apart one day when a herd of young heifers was put into the paddock that we passed by with the bulls every day. They came galloping up to the fence to inspect the bulls. What pandemonium there was, as the bulls started bellowing and snorting and jumping about. I enjoyed it very much and laughed every time we went by, but Bonefacio was furious and had the heifers moved into another field. I was sorry about that for I enjoyed the morning walks with the bulls watching the heifers getting excited as we approached the fence. This they did by jumping about and throwing up their heels and squealing.

This took up all the morning. The bulls were then brought back to the galpon and were fed with freshly cut alfalfa. We then had our lunch followed by a siesta for ourselves and the bulls. Again we had to muck out and add fresh bedding and if any bull had slept on its dung, he had to have his backside washed. They were treated just like newborn babies. More feeding and watering followed, till the sun went down. But our day was still not finished, for the three of us would return to the galpon and remove any further traces of dung and give their bedding a final shake and feed them with hay. By 9.30 I was fast asleep. On Sundays, the only difference was that we did not have to exercise them.

I continued in this job for six months and, by then, I could speak Spanish, more or less. I was still homesick but refrained from saying anything to my family or Sir Arthur. I did have one treat and that was on Sunday afternoons, when Mr Simpson asked me over to have tea with his wife and family, mainly, I think, because they realized that I was determined to stick it out. Also, I became aware that Bonefacio treated me, more or less, as an equal. In fact, he told me that he never ever expected to have an Englishman under him. This change by Bonefacio made life for me more bearable, and, of course, we could now converse with one another.

At this time, I became aware that Bonefacio was at odds with the peon who milked the cows that supplied milk for the Simpson family and on several occasions I had heard the two of them cursing one another because, as the cows passed our front door twice a day, they deposited their cow pats as 'visiting cards'. One day I was reading a letter from my mother, when I became aware of shouting and swearing between Bonefacio and the milk man. At first, I took no notice, as you often heard this sort of swearing, but, after a minute, I could hear the 'click-clack' of what I knew at once to be the sound of knives meeting one another. Both men were good fighters and pranced around each other on the tips of their toes. Each was aiming blows with his knife at the other's forehead so that the blood would temporarily blind him, and then he could finish the fight by plunging his knife into the other's stomach. So far neither had succeeded. The whole fight suddenly changed because Bonefacio slipped on a fresh cow pat, (the whole cause of the fight) and fell on his back, losing his knife at the same time. This was an opportunity not to be missed by the cowman and he threw himself on to Bonefacio preventing him from getting up or reaching for his knife. The only chance for Bonefacio was to grab the wrist of the cowman to prevent the knife entering his stomach, exerting all his strength in keeping the knife at bay. Little by little the cowman was getting the point nearer and nearer to my friend's stomach. I was completely transfixed at the door as I had never seen a fight of this description before. The shouting and the click of knives had alerted the other men who rushed to the scene. The Spaniard who worked with me, saved Bonefacio by kicking the arm of the cowman, deflecting the point of the knife, which skidded round Bonefacio's ribs, cutting the flesh but doing no real damage. The cowman was dragged away and Bonefacio got to his feet with the blood flowing from his long cut. Staffer soon arrived and the cowman was locked up until the police arrived and we never saw him again. Meanwhile, I had the job of bandaging Bonefacio's wound every morning, but he carried on with his work. Although this has taken some time to set out, everything happened within two minutes. My reactions were non-existent. I suppose if I had been a peon, I could have acted quicker.

Soon after this, I was told by Staffer that I would be able to work with the Capatas of the farm, who was responsible for all the mounted peons who looked after about 5,000 Aberdeen Angus cattle. This was a great step up for me, as the men who worked with cattle were a race apart from the foot peons. These men understood everything concerning cattle; they were excellent riders, they knew how to lasso cattle, skin dead animals, part out stock, mend fences and many other jobs. Another advantage was that Sunday was free unless there was an emergency such as a fire during the summer, and the cattle were in danger. I had learnt only a little riding. In fact, my only experience of riding was when I went to the village to cash my salary cheque at the end of the month.

This was always a great event for all. The peons would saddle their horses, put on their best clothes and make sure they changed their everyday knives for the really sharp ones, in case they ran into trouble. About twenty of us would dash to the village in a cloud of dust, shouting and yelling at the tops of our voices and the first stop was the local brothel which was situated on the outskirts of the village. Here the 'girls' would be waving and inviting us to stop and have drinks. I was the only one who continued into the village to change my cheque. After having my hair cut and buying little luxuries, I would ride back, passing the brothel; no sign of the men – just a row of tied up horses fast asleep.

It was not long before the peons told the 'girls' that there was a young Englishman working with them and the following month I was greeted from the windows by the 'girls' inviting me to get down, come in and drink maté and dance with them. I was too green and frightened to accept their invitations, but they never failed to wave to me as I galloped by.

There were about a dozen men who worked the cattle and, of course, the Capatas gave the orders. He in his turn would go to the office each evening to talk to Simpson and Staffer to discuss the work to be done the next day. My gang was delighted to have me with them for they knew they would have plenty of fun watching me fall off my horse. I was given a string of horses for my use which was brought into the corral with all the others twice a day and I picked out one in the morning and another in the afternoon, as chasing cattle was tiring for horses. My horses were well trained for cattle work and knew exactly how to part out cattle. I think the first day I fell off my horse five times, but I was only bruised. The men would shout out to me to take out from the herd a certain cow ('parting out'). I would then chase after it with the horse at full stretch – but unfortunately when the cow made a sharp right-angled turn, the horse would do the same, but I would not. I would go straight on and find myself on the grass! This was just what the peons wanted to see me do – and me an Englishman to boot. How they laughed! Doing this work all day long from sunrise to sunset soon taught me how to ride. Also I learnt how to skin dead cattle – I carried my knife in my belt as they all did – how to mend fences, to oil windmills and many other things. This kind of work I did for three months and then I was brought back into the galpon to help Bonefacio in preparing the bulls and heifers for the next Palermo Show.

* * *

It was about a week before the Show, when the pedigree bulls were coming to perfection, that I was asked to go to B.A. with Bonefacio. This was very interesting as I saw all the best breeds in the Argentine on display and, of course, I was back in civilization again and even had time to see my friends. In the Aberdeen Angus section we got two second-class prizes for bulls of different ages, and two more recommended. After the Show, I went back to work with the cattle, but after two months, an epidemic of foot and mouth disease broke out.

In the Argentine, at that time, there was no policy of slaughtering cattle as we have in the UK. We just remained isolated until the disease abated, leaving the cattle thin, killing off the old cows and the very young calves for they were unable to suck from their mothers. Once the disease had passed, the cattle would start to put on weight again but it would take a year, at least, to get them back to a condition in which they could be sold. Not all the cattle would get the disease. Sometimes the virus would miss two paddocks of cattle and start up again amongst cattle three miles away. In time, the disease broke out at the galpon among the bulls that had been in B.A. and I was called back to help Bonefacio cure them. The only method in use at that time was to wash their gums twice a day with permanganate of potash and an ointment rubbed well in between their toes. They were fed only on freshly cut alfalfa. After two weeks, the bulls responded to the treatment and the dreadful dribbling stopped. Unfortunately, I became infected with the disease myself. My gums became very sore indeed so that I was reduced to taking only soup and bread and milk. I had to wear a handkerchief under my chin because I slobbered like a baby cutting teeth. I, too, used the permanganate of potash on my gums. I lost a little weight but my feet were not affected. When I returned to England, nobody, especially veterinary surgeons, would believe that I had had foot and mouth disease – humans don't get it, I was told – but years later it was admitted that humans *can* catch it. It was just as well I was in the Argentine, otherwise I might have been slaughtered and burnt!

One evening, having worked with the cattle all day, branding, castrating, ear clipping and de-horning, we were sitting round a big fire, sucking maté, exchanging amusing stories of what had happened during the day, the older members recounting the adventures they had had during their youth, when we became aware of a man approaching on horseback, with a black dog and leading a spare mare. He rode up to the hitching rail and dismounted, greeted us in the usual way, "Buenos noches, Senores," and inquired if there was any work for him with the cattle at the estancia. I noticed that his lasso was neatly tied to his saddle and that he was of a type quite different from our local men. He was dressed in top boots with large pointed spurs that jingled as he walked towards us, but what struck me most was the belt he wore which was studded with silver coins and which had a large silver-handled knife stuck into it. By his dark, straight hair, high cheekbones and very dark eyes, I should say he had a sprinkling of Indian blood. He certainly was not from our part of the country, but in all probability from Corientes, in the north of the Argentine. His horse's bridle had plenty of silver bands on it and one or two on the halter.

Our estancia was completely enclosed by a very good fence indeed, not only to keep the cattle in, but also to keep other people out. There were only two entrances, one to the south and one to the north, with two roads leading to the estancia and Mr Simpson's house. At the two entrances were two houses inhabited by peons and their families. These peons were responsible for the cattle in that area, but above all, they had to see that no one left the estancia to visit the village after sunset, without a pass with his name on it, signed by Staffer. The gate was locked at sunset and not opened till sunrise. During the day, anybody was free to come to the estancia looking for work. There was, in my day, an old custom that if a man could get into the camp before the gate was locked at sunset he was entitled to spend the night with our peons, his horse would be released amongst our horses, and he would be given a meal with us. In the morning, he would ask the major-domo if there was any work and if the answer was "No", he had to leave the estancia at once.

We invited the stranger in, as was the custom, and I noticed that my fellow peons were very interested in him but a little apprehensive.

Next morning, as the sun was peeping over the flat land, Staffer came down from the estancia house to give orders for the day's work. As we were one man short and the newcomer said he was experienced with cattle, Staffer took him on. He was certainly one of the best men I have ever seen working with cattle and was even better than our own Capatas de Campo.

After a week, I began to notice that many of the peons tended to keep away from this new man. On making inquiries, I was able to find out that he had only been released from prison a month ago, having killed a man with his knife in a fight. He was known as Videla – a short thick-set man of about thirty-five, unmarried, and a superb rider.

If you recall, before my journey out to the Argentine, I mentioned that when my mother said goodbye to me, she gave me four five-shilling pieces to take with me which, she said, would bring me back safely. (These four five-shilling pieces were given to her by her aunt when she left with my father for the Argentine after they were married.) One day when there was a swell and we were just sailing off the Azores, sitting on the boat deck with three or four people, the conversation turned towards coins, as one of the men had a collection of them. I mentioned my five-shilling pieces and fetched one from my cabin. One of the ladies asked me if she could examine it. She then handed it to another lady who, unfortunately, let it drop on the deck and before anyone could rescue it, it had rolled into the sea. I was very sorry about this and now I only had three left.

I now return to the estancia, and Videla and the story of the other five-shilling pieces. It was the summer when the growth of the thistles made it imperative that we kept a close watch on the cattle. These thistles grew to a height of six feet but, unlike Scottish thistles, did not carry any spikes. During the heat of the day the cattle sought their shade. The thistles grew in a large paddock of about 1,000 acres, dead flat, no trees beyond a few weeping willows planted close to the water troughs. The fences stretched in a straight line as far as the eye could see. Our duties were to inspect the fences, check the windmill, see that the Australian holding-tank was full of water and, above all, examine the cattle, all Aberdeen Angus, as they were recovering from an attack of foot and mouth disease. It was most important to find any dead animals and skin them before the hides became useless through putrification. This particular herd of 300 was the responsibility of Videla and myself. Most days we would find one or perhaps two dead animals. Working together, Videla and I soon became close friends and he taught me many things about working cattle, lassoing, skinning dead animals, mending fences and many other things. One day I asked him why he had killed a man. He told me it had happened up north in a boliche (pub) where both men had been drinking. A quarrel arose and angry exchanges reached such a pitch that the other man shouted at Videla the biggest insult he could: "Go back to the whore that gave you birth." You may call a man a bastard, but you must not directly insult his mother. Knives were immediately drawn and the fight was on. In a few seconds the man was dead and Videla was taken away by the police.

Having mentioned Videla's belt with the silver coins, which I much admired, it was not long before I showed him my five-shilling pieces, after which, he was always asking me to give them to him. One day, after inspecting the cattle, we dismounted under the weeping willows, tied our horses to the fence and rolled our cigarettes. Videla suddenly asked me if I knew how to fight. I told him that at school I had learnt how to box. He roared with laughter, drew his knife and asked me if I could defend myself. Of course I could not. "Well," he said, "I will teach you how the Gauchos fight with knives – you never know what might happen." So I agreed that he should teach me and he said he would, on one condition; that I give him two of my five-shilling coins. He then gave me a demonstration with his knife and I was amazed at the dexterity he showed. He danced round me on his toes, as light as a feather. Next day, he cut out two wooden knives and I had my first lesson, which was how to hold a knife with the tumb well tucked under the fingers so that it was guarded, for if your opponent cut your thumb, you could not hold your knife. After about four lessons, I got the hang of it and, in some measure, I could defend myself. I then gave him the two coins and Videla slit two holes in his belt on either side of the buckle and inserted the coins. How proud he was and the envy of all the peons.

Some days after, disaster struck. Videla and I had spent all the morning in our paddock revising the stock. Amongst the thistles, we found two dead cows, and we promptly skinned them. While I was skinning my cow, which was quite close to Videla's, I heard him swearing so went over to him to see what the matter was. It seems that while skirting round the cow's hocks, he broke about an inch off the point of his knife. He asked me to lend him mine to finish off both the hides. Having done this, we rode back to the estancia with the hides behind our saddles and took them to the Saledero for salting. After lunch, we took a siesta till three o'clock when the bell sounded for us to collect fresh horses for the afternoon. All the cattlemen, including Videla and I, had saddled up and were waiting for Staffer to arrive with our instructions. Most of the men were sitting on their heels, their favourite position, smoking. I noticed that Videla had not changed his knife and it was in its scabbard behind his back.

At last Staffer came out of his office and, in a loud voice, demanded that Videla should come to him at once. I could see, by Staffer's face, that he was furious. Videla rose from his heels and went towards him. Suddenly everybody became tense, for Staffer was losing his temper and calling Videla every name under the sun. It seemed that after we had left our paddock that morning, Staffer had gone to inspect the cattle and had found two more dead cows in the thistles. They had been dead some time and their hides were useless. Videla started to defend himself saying that he could not search every inch of the thistles. This only infuriated Staffer more and he completely lost his temper and called Videla a murderer and, worse still, he made the fatal mistake of saying, "Go back to the whore that gave you birth." At these words, all the peons and myself became frozen, for we knew that Videla would not accept that sort of insult. He took one step forward and Staffer realized that he was in real danger of being knifed. We all knew that Staffer had a small automatic revolver in his hip pocket and I also knew that Videla would not draw his knife for the simple reason that the point had been broken that morning. He knew that he could do little damage with it and this would give Staffer a chance to draw his revolver and shoot him. Meanwhile all the peons expected Videla to plunge his knife into Staffer's stomach, instead of which, Staffer, realizing his danger, had taken a step backward, so as to be able to draw his revolver. Videla too saw his danger and grabbed Staffer's tie and at the same time his hand went to the handle of his knife, shouting, "If you draw your gun, I shall knife you." Staffer had regained his senses and he knew he would be a dead man if he made any attempt to draw his revolver.

All this time, everybody had been rooted to the ground as this drama was taking place, expecting any moment that Videla would kill Staffer. I was the only person there among all those men who knew the real position – Videla had won by a colossal bluff. Staffer retired back into his office, thought things over and, realizing his terrible mistake, came out and told Videla that he must leave the estancia at once. So, after collecting his other horse and dog, he rode up to me to say "Goodbye" and wished me well. He shook my hand, and as he did so I saw the setting sun shining brightly on his silver belt and my two silver coins. I wonder what happened to him and what became of the two coins, depicting St. George on his horse killing the dragon and Queen Victoria on the reverse. My knife with its four-inch blade, I have to this day. What a drama that was on that hot summer's afternoon – never to be forgotten. I found out afterwards that Staffer suffered from stomach ulcers which probably accounted for his bad tempers.

Within a short time, I went down with jaundice and had to go to B.A. for treatment. The doctor advised me to return home, which I did, in September 1923.

* * *

Sir Arthur was away in Crete so I spent my 21st birthday at Blagrove. Only my sister Ethel was at home but she made me a lovely omelette for my supper!

What a lovely feeling to be back in England again and with my family at Blagrove. For a whole month I revelled in doing just nothing with not a care in the world. La Isabel and the peons seemed to just fade away.

When Sir Arthur returned to Youlbury, I was on a different footing with him. No longer was I a young teenager – I could come and go as I pleased. Sir Arthur was delighted to see me, and Youlbury again became my home. Dear Ada was so happy that she could look after me as she did in the old days. The shooting season was in full swing and I was invited to join in several shoots, not only by my neighbours, but also Mr Humphrey of Upton, and how that reminded me of the days when I first went there after leaving Trent College, and of my great friend, Frank Madge. While I was in the Argentine, he had married Woggie Beal and already had a daughter. It was not long before I was staying with him on his small farm at Horley in Surrey.

I told Sir Arthur how I had cured the bulls of foot and mouth disease and what had happened to me. He suggested that I should write a full description of my treatment of the disease and he would send it to the Ministry of Agriculture as an alternative to the slaughter policy here in England, but the Ministry said they would continue their policy of slaughtering.

One night after dinner, and a game of billiards, which Sir Arthur always enjoyed, the conversation came round to what I had in mind for my future life. I had given this question a great deal of thought, for I certainly did not want to join my father and Martin in farming at Blagrove. I felt I could not throw away all the experience I had gained in the Argentine, especially all I had gone through at La Isabel in order to learn the language. Also, I wanted to prove to Sir Arthur that I was not a failure. When I told him that I was determined to return to the Argentine, he fully agreed with me. I had already approached my uncle asking him when he would be shipping sheep and bulls to B.A.

In January 1924 I sailed from Cardiff on the *Oakland Grange*, an old cargo boat, with a dozen pedigree Lincoln rams. The only accommodation was astern where, during the Great War, the gun crews' quarters had been. This time the weather was kind.

After passing the Equator, one of the seamen was painting the outside of the bridge, when he fell on to the deck below, breaking his leg in two places, and the bone was sticking out through his flesh. As we had no doctor on board, we sent an SOS to any ship that might be carrying a doctor, and it was not long before the Italian passenger liner, *Princess Mafaldo* turned up. Meanwhile the seaman had been bound up in a stretcher. Both ships had now stopped, a boat was lowered from the passenger ship and a doctor came aboard. It was at that point that I became aware that circling round the little boat from the liner, was a number of large sharks and when the injured man was lowered over the side down to the little boat, the sharks became very excited because of the blood which had seeped through the bandages. The sharks almost jumped out of the water and the crew had to jab the fish in their sides with their boat hooks. Once the man was safely in the boat, they rowed off to the passenger ship and we continued on our way.

Some days after this, the engines broke down and we stopped for repairs. The next morning I found one of the rams dead in its pen due, I think, to its long heavy wool and the high temperature. After breakfast, we threw the dead sheep overboard and all the crew watched to see what the sharks would make of it. They had the devil of a job to dispose of the carcase because all they got at first were mouthfuls of wool which got tangled up in their teeth.

Again we called at Montevideo to discharge cargo and the rams became very restless for they could smell land once more. As soon as I had cleaned and fed the sheep, I changed my clothes, for I, too, wanted to feel solid earth beneath my feet again. We were berthed in a very run-down part of the docks with rather mean-looking houses. As I walked down the gangway, I noticed a boy of about twelve years of age and as I stepped on to the road, he handed me a card. I naturally thought that it was an advertisement for an hotel or restaurant, but when I saw it, I realized that I was mistaken, for all it said, was 'Lulu, Calle (Street) Paradiso 14'. When I looked at the boy in a rather surprised way, he immediately said, "My sister very good 'jig-jig', pink inside, like English girl." I was absolutely horrified and dashed back up the gangway and showed it to the Captain who roared with laughter and called the Chief Engineer and showed him the card. They turned to me and said, "Jim, we think it is time you were 'educated' about what goes on in the world. We shall take you out tonight and call on this young lady and we can assure you you will be perfectly safe." At first, I refused, then my curiosity got the better of me.

On arriving at Paradiso, I noticed that all the doors were closed, but each door had a peephole and when you slipped the covering aside you could see what was going on inside. I also noticed that there were other men in the street taking advantage of this method of observation. At first I thought they were extremely ill-mannered to look through people's peepholes, but on arriving at No. 14 the Captain invited me to have a look, which I did with some misgivings, and there I saw a patio furnished with chairs and tables. There was a small area in which to dance and a bar at the back. Sitting on the chairs were three or four scantily-dressed girls. I realized that this was a brothel. What would my mother or Sir Arthur have said? Well, there's nothing like 'learning', so in we went and sat down, to the great delight of the girls. The Captain ordered champagne and when the girls brought the drinks, they sat down on our laps. We chatted away and drank the champagne. Suddenly my girl planted a kiss on my cheek. This was too much for me. I promptly got up and dashed through the door, up the street and away to the ship into the safety of my cabin where I grabbed hold of a large bar of carbolic soap and washed my face! I did get my leg pulled when the Captain and the Chief Engineer returned, having followed me back.

On arriving at B.A. we were met again by the same flea-bitten old ewe that I had seen on my previous trip. Again the eleven pedigree Lincoln rams meekly followed the ewe and were placed in quarantine for thirty days. After the rams came out of quarantine, they were moved to the showrooms. A. Bullrick, the auctioneers, had very skilful men to look after the sheep to get them back to their prime condition, especially their fleeces which were matted due to the salt spray from the sea. The condition of their wool was a vital factor and much time was taken to bring the fleeces up to show standard. At last the day of the sale arrived, bringing buyers from all parts of the Argentine, and even from Patagonia, as my uncle was well known for exporting good pedigree sheep and bulls. My eleven rams sold well and my uncle gave me a bonus for looking after them on the journey from England.

* * *

1924 was a most important year in my life, never to be forgotten, for in March I met, through friends, the girl who was to become my wife. It was through Mr and Mrs Hansen, with whom I was staying at the time, that I first met Kitty Cork, at the local cinema in Lomas, a suburb of B.A.

When we came out of the cinema we all went to the Hansens for tea. I must say I was instantly attracted to Kitty. She was young, good-looking and had a slim figure and for the first time in my life I felt at ease in a girl's company. While we were having tea, a thunderstorm broke out so I ordered a taxi and took her back to her home. I managed to screw up the courage to ask her if she would have tea with me in B.A. the following day, after she had finished her typing and shorthand lessons. From that moment I saw as much of Kitty as I could. The Hansens saw that I was interested and very kindly asked me to tea and to stay the night on several occasions. I remember taking her to the cinema in Temperley and actually, with great daring, I kissed her, while we were talking on the doorstep of her house and then took to my heels without even saying good-night. The next day I telephoned her and apologized. Her answer to me was, "I liked it." From then on our friendship blossomed.

Kitty had been born in the Argentine in 1906. Her father was a Quantity Surveyor for the Great Southern Railway which, at that time, was owned by the British.

I went round the agencies looking for a job on one of the English estancias, bearing in mind that I had to have an outdoor life because of my ear. I had offers from English and American banks but I turned them down. One day I received a letter from Krabbe, King & Co., who were agents for many English estancias, informing me that a certain Charles Jewell was looking for an assistant for his estancia in the province of Santa Fe. I called at their offices and was told that the salary commenced at 150 pesos per month, all found. Mr Jewell had recently arrived from England and was staying at the Plaza Hotel and I was to go there for my interview. I dressed in my best suit and wore my spats! In fact, I have them to this day – sixty years later!

I was not very impressed with Captain Jewell, as he liked to be called – a pink, bloated gentleman who wore spectacles and, as it turned out, I never liked him. He offered me the job at his estancia which was called Las Petacas. I telephoned Kitty and asked her out to dinner to celebrate. The following Sunday I was introduced to the Cork family in Lomas. Mr Cork was a very strict man in the typical Victorian tradition. No daughter of his could go out with a young man who had no job. Mrs Cork was not cast in the same mould, but she was very careful not to cross her husband. I think she knew that Kitty had been going out with me but her husband did not. Now that Kitty had a young man who worked out on an English estancia, the situation was altered and I became quite a frequent visitor until I left for Las Petacas. The rest of the family consisted of a married daughter called Margaret, Stephen a bachelor, working in an American bank, and a younger son, Jack, an engineer on the Southern Railway.

With what joy I wrote to my parents and Sir Arthur to say I had a real job and good prospects. I knew Sir Arthur would be more than pleased and I felt that I had a real chance to show him I could stand on my own feet.

On arriving at San Jorge station, I was met by the secundo major-domo and received as an equal; very different from my reception at La Isabel. Las Petacas consisted of about thirteen leagues, a large part of which was rented to colonists, divided up into one hundred-and-fifty hectares to each family. These colonists were mostly Italians but there were also a few Spaniards and they paid their rent based on a percentage of their total harvest. The crops mostly grown were wheat, linseed and maize. The rest of the estancia was devoted to Hereford cattle and certain paddocks were planted with cereals.

I was engaged by Jewell as an assistant to Oliver Cassels, Jewell's son-in-law, who was in full charge of the Colonia Santa Anita which was part of the estancia. This portion of the estancia was very important as we received a vast tonnage of cereals in lieu of rent, which was collected and stored in a very large galpon and then sold for export to Europe. The overall management of Las Petacas and the Colonia Santa Anita was in the hands of the major-domo, James Wood, a bachelor, who had his own house and lived apart from the rest of the staff, which consisted of the secundo major-domo, Field, the secretary/accountant, Frank Corkery and two other young Englishmen responsible for other departments. I was known by everybody as Don Santiago and I was never addressed, even down in the town of San Jorge, as Mr Candy.

Cassels and I were entirely responsible for the Colonia Santa Anita. Only occasionally were we called upon to help with the stock, during the rodeo when the marking and branding, etc. took place which was during the winter. We bachelors all lived together in the same house. We had separate bedrooms, a communal sitting-room and dining-room, a kitchen and two maids to cook and look after us. We had electric power produced on the estancia, but we lacked proper toilet facilities. There was only one WC, the 'Long Drop' which was situated some distance from our quarters in a small paddock. It had a tin roof and was guarded by a pedigree Hereford bull. What an obstacle that was! There was no refrigeration; we had ice blocks delivered twice a week when the coach and horses went down to San Jorge for the mail and stores.

The estancia house had an interesting history. It was built before 1850 and the old part still remained in the form of a square. But over the years it had been enlarged to make accommodation for guests and servants. On the flat roof there were two small brass cannon, not for defence, but for use when the owner in those days received news that the local indigenous Indians were attacking the stock. Then, having gathered his staff together, they would set off to hunt down the Indians, and, if possible, kill them.

When the country was first opened up by the Europeans and cattle were introduced, they were plagued to death with the Indians stealing them. This state of affairs was not to be tolerated, so by degrees they were killed or driven into areas where, in time, they died out, mainly from starvation or disease.

At Patacas there was on the boundary a large low flat area called the Cañarda de San Antonia. This area had dense tall grass and reeds which gave good cover for the Indians. Naturally the Indians became very fond of eating the cattle. This had to be discouraged, but it was difficult to catch them in the open paddocks as they took refuge in the vast area of tall reeds. The major-domo at that time, an Englishman, decided to wait until the winter when the reeds were dead and dry and to set fire to the whole area, and if any Indian broke cover, to ride them down and shoot them. One day, after a very successful hunt, one of the peons who was hunting them, was passing a small pool of water with green rushes growing round it, when he heard a baby crying. In the rushes he spotted a woven hand-made basket and in this basket was a new-born baby. This peon's wife had recently lost her baby, so he picked up the crying infant and took it back to the estancia and handed it over to his wife who brought it up as her own child. Here again is the story of Moses repeated. This child, Juan Tolosa, grew up at Petacas and became a very good cattle man in the best tradition, worked all his life with cattle, married a Spanish girl and, in time, was given the 'Puerto Norte' (the north gate entrance house) and was responsible for looking after the gate. In 1924, when I arrived, he had retired and his eldest son had taken on the job and Juan spent most days sitting in the sun talking about the past. He must have been the only full-blooded Indian I had ever seen in that part of the country. At that time he was well into his eighties or may even have been ninety. When my father was in the Argentine in 1885, the local Indians had already disappeared. In Tasmania the Europeans had the same trouble, but they got rid of the natives by driving them over the cliffs into the sea.

I must now return to my story. Las Petacas was a self-contained village; the estancia house, James Wood's house, our quarters, the butcher's shop, a general store with food, clothing, cigarettes, etc., blacksmith's and carpenter's shops, the peons' quarters, corrals, galpons, etc. It was paradise to me after the Isabel. But we were completely cut off from the rest of the country so that we never had visitors and I never saw an English girl for two years except once a year when we had a few days' holiday in B.A. Our weekends consisted of playing tennis, shooting, playing poker and getting drunk. Even when Jewell came out, bringing guests with him, we kept our distance for none of us liked him. Things did change when Wood married an English girl from B.A. and she invited her friends to spend holidays with them. On my annual visit to B.A. I saw as much of Kitty as I could and we became very good friends but it always took a few days to get into the swing of life in B.A. after the months of complete quiet and isolation of Las Petacas.

1926 was a momentous year for me. Kitty had decided to take her holidays in the Cordoba Hills, about two hundred miles from Las Petacas. I went and saw Wood and he agreed to give me ten days' holiday. We were to stay in a nice hotel called Los Montes, which was situated by a small rocky stream and was run by an Englishwoman. When Kitty told her mother that I, too, would be staying in the same hotel, she insisted on Kitty having a chaperone! Also at the hotel were a number of other boys and girls, but no chaperones for them.

One day, we all decided that we would take a picnic to a wonderful place called Fuerto Mala which was about three miles from the hotel. We had changed into our bathing things before leaving the hotel. We arrived at the waterfall which fell into a large pool from a height of twenty-five feet. The pool itself below the fall was very deep, but shallower round the edges.

It was not long before we were all swimming and thoroughly enjoying ourselves, that is, all except Kitty as she could not swim. For this holiday she had bought the latest Jansen swim-suit without a short skirt, and a dark blue bathing cap. After a time we came out to rest on the rocks, except Kitty who was splashing about in the shallow part. I don't think anyone had realized that she could not swim; they all thought because of her modern bathing dress that she could. Suddenly, I heard a shout and looked out into the pool. There, to my amazement, was Kitty in the deep water, then I saw her disappear beneath the surface. I jumped to my feet and dashed into the water and, to my horror, I saw her disappear for the second time just in front of me. I could see her blue cap under the water. I dived straight down and managed to grab her arm and pull her to the surface. By this time, everyone had jumped into the pool. We got her out, turned her upside down so that any water could escape from her lungs and then laid her on the bank and carried out the usual drill and it was not long before she was sitting up again. The girls took her wet bathing costume off, dressed her and wrapped her in towels and sat her in the sun where she slept for over an hour, after which she was her old self again. I, of course, became the hero of the day!

After dinner that night, I asked Kitty if she would like to go for a walk with me. She said she would be delighted. I remember when we were clear of the hotel we stopped under a tree and stood looking at one another. Then we fell into each other's arms and kissed. We then realized that we loved one another. Kitty kept saying that if it had not been for me, she would now be dead. When we arrived back at the hotel, the girls gave us some knowing looks. Next day I had to leave for Petacas but Kitty still had three more days' holiday. She came to the station to see me off and as I leant out of the train,

I could read from the expression in her eyes that she loved me.

I had to change trains at Cordoba City which meant that I had to stay the night there before catching another train to San Fransisco where the estancia car would pick me up. That evening I had a look around Cordoba City and inspected the ruins which had been caused by an earthquake.

My room was a corner one and from the corner of the building there was a wire connected to a pylon from which the local trams received their current. I had fallen asleep when a train came trundling along and the sound came loud and clear through the wire into my ear and, in a second, I was out of bed, down the stairs and out into the street in my pyjamas, much to the amazement of the janitor. I was quite sure that it was an earthquake.

When I arrived back at Petacas, I fully realized that I had fallen in love with Kitty. I think the real understanding came when I pulled her out of the pool. Now, I had been brought up to believe that if you kissed a girl, it was because you were in love with her and you wished to marry her. How strange all this sounds in 1984. I was very perturbed for, with my present salary of one hundred and seventy-five pesos per month, I could not even get engaged. So I sat down and wrote to Kitty and told her that although I loved her dearly, I could not ask her to become engaged to me. To soften the blow, I bought her a very nice camera and asked her if I could write to her as often as possible. She replied that she fully understood my position and that she felt the same way as I did. Again she thanked me for saving her life and hoped that I would write as often as possible. From then on, letters were our only means of communication and what a joy they were for us both.

Meanwhile, I had been given full charge of the Colonia Santa Anita and my salary had been increased to three hundred pesos. At that time the pound was worth thirteen pesos. Today it is nineteen thousand, eight hundred pesos. The whole of my time was devoted to seeing that the colonists carried out their duties – cultivation, harvesting, etc., in accordance with their tenancy agreements.

As I mentioned before, the only time I helped with the stock was during the rodeo. This lasted about three weeks when all the calves that had been born during the year, were rounded up, lassoed, branded, de-horned, vaccinated, ear clipped (to denote year of birth) and the bull calves castrated. All the cows and calves were penned in between two right angles of a fence with men on horses to prevent them from breaking out. This operation needed all the men and boys possible. First, a large fire had to be lit and kept hot, and the boys had to run back and forth with the branding irons. We assistants were each responsible for the various important jobs that had to be done to each calf.

When the branding irons were red hot, the signal was given to the peons on horseback to move quietly into the herd and lasso the calves round the neck. The mothers would then start bellowing loud enough to almost deafen you. The lassoed calf was dragged out of the herd by the horse, dashing about like a fish on a hook, but it could never get away as the lasso was attached to the saddle. At the same time, there were several peons on foot who were expert at throwing the lasso just in front of the galloping calf's legs. The object of the throw was to jerk back the lasso loup when the two front legs of the calf had dropped into the ring of the lasso. The lasso had to have a three feet circular coil on the ground and the throw so timed that it landed just in front of the calf. Then, at the right moment, a slight jerk by the peon and the coil would bring the two front feet of the calf together inside the ring and the calf would just pitch head over heels. The thrower had to keep his lasso taut to prevent the calf getting up again. Meanwhile another peon would sit on the calf's head at the same time loosening the lasso so that the rider could use it for the next animal.

Then came the following routine; first, the red hot iron with our mark 'LP' on it was brought from the fire at top speed by the boys and stamped on the top part of the hind leg. The branding was done by the secundo major-domo. This had to be done carefully. Then the vaccination, the ear marking and the tally indicating the sex of the calf. Next came the de-horning which was done by me with a very sharp knife, or a two-handled pruner, if the calf was over nine months old. This was a nasty job because they would bleed, so a caustic stick was applied which stopped the bleeding and prevented the horn from growing again. The most important job was castrating and that was done expertly by our major-domo, James Wood. The noise was terrific and the dust rose from the herd in thick clouds. We were better off as we worked away from the herd, unless the wind was blowing in our direction. Among the peons, testicles from the young bull calves were considered a great delicacy to eat. When they were removed, there was a boy holding a tin in which to receive them and then they were threaded on a skewer and barbequed on the fire.

At midday we stopped to have our 'Asado', which was cooked on the red hot embers. All we had to do was to hack off a piece of meat and wash it down with a little wine followed by maté. There were times when we would cook a calf which had to be killed because, when being lassoed, it had fallen badly and broken a leg, and very good it was too.

On one occasion, I was persuaded to try the testicles cooked on the fire. They produced a fine bull calf and, as I did the de-horning, I noticed the calf was 'wall-eyed', rare in cattle. Within a few minutes the boy brought back from the fire this particular calf's testicles. My first reaction was not to eat it, but then I thought I must do as the men did as they and the boy were watching. Well, I found them very good indeed.

About an hour later, when very few calves were left unmarked, I was feeling very tired and sitting on an orange box, just a few yards from all the dust stirred up by the herd, when suddenly out of the dust rushed the bull calf whose testicles were, at that very moment, in my stomach. Before I could even get off the box, this wall-eyed calf hit me in the chest, knocking me flat on my back. What is more, his hind hoof cut into the calf of my leg leaving a scar that I have to this day. I am sure that the calf must have derived some satisfaction from seeing me sprawling in the dust.

Life went on in its monotonous way, month after month; nothing to look forward to except the twice weekly incoming mail. This was an open coach pulled by four horses that took a whole day to go to San Jorge and back, leaving the estancia early in the morning and generally arriving back with the mail and general stores and 'galleta' (a very hard biscuit that would keep for months). This biscuit was for the peons. We had bread which came with our two precious blocks of ice and several cases of beer. Frank Corkery and I used to sit out on the veranda of the office to await the arrival of the coach. Frank's first duty was to receive the large leather bag from our coachman, unlock the padlock and shoot the mail on to the office table. I was, of course, looking for Kitty's letter, for by now she was writing once a week to me and what a joy those letters were as they drew us both closer to one another. Also, from time to time, my mother and Sir Arthur wrote to me.

After the harvest, which finished at the end of February, I took my holidays and we motored down to Mar del Plata, south of B.A. One of the best seaside resorts, Mar del Plata is much frequented, during the season by rich Argentine families. It has every amenity you could think of. Not only has it fine sands but when the tide comes in, the breakers are large and if they do knock you down, you are very quickly carried back up the beach, causing screams of delight.

It is well known that pounding water is good for reducing weight so when the tide comes in, the ladies with large bottoms, all join hands and wade into the sea till the water is up to their knees. When they see the breakers coming, they all turn round and stick their bottoms towards the waves which crash down hitting them with considerable force. I, being slim, much preferred to dive through the crest of the wave, turn round and see what had happened to the line of fat women. Usually they were swept up on the beach but they always went back for more.

Our party consisted of Mrs Hansen, Maud, Ernest and his wife and, of course, Kitty. What a wonderful time we had together, bathing shopping and dining in nice restaurants. One day we were all sunbathing, when Ernest thought it would be a good joke to bury Kitty in the sand. This we did, till only her head and shoulders were to be seen. Just as I was scooping up the last handful, my fingers came into contact with a thin chain. On pulling it out, I was amazed to see that the chain was attached to a beautiful slim gold watch with scrolls round the face and on the back and, when wound, ticked away as good as new. It must have been there for some time for it was about a foot down in the sand. As there was nobody close to us, I put the watch in my pocket and said nothing.

Alas! the day came when we had to return home. Having filled the car with petrol, we counted up the money we had left and I was the only one that had any – a one peso bill! As we passed the Casino, I had a brilliant idea of going into the Casino to the nearest roulette wheel. We pulled into the side and as I leapt out of the car, I told them I would be back in a few minutes. I joined the crowd round the table and placed my one peso bill on No 9 black and stood back to watch the ball spinning round and, to my utter astonishment, 9 black turned up. The croupier pushed towards me, with his rake, thirty-seven white chips which I hurriedly gathered up and changed for thirty-seven pesos. With the money in my hand, I jumped back into the car and you can imagine the surprise of the others and the roars of laughter that I had won enough money to pay my expenses in Mar del Plata. The Hansens wanted me to go back and gamble again, for they were sure that I would win a fortune. This I refused to do and we set off for home. I often wonder what would have happened if I had continued to gamble. The crowd round the table had gazed in astonishment when I rushed away. Kitty and I always looked on No 9 as our lucky number, when buying raffle tickets, etc.

It was after this holiday that I knew I wanted to get engaged to Kitty. In a way it was a bit of a shock as I had never been in love before. When I returned to Patacas, I realized that I would have to do something about it. My first move was to consult my great friend Frank Corkery. His advice to me was that I should ask Kitty to become engaged to me. At that time at Patacas not one of the staff was married, not even the major-domo, and he had a three-bedroomed bungalow with all mod cons! If I wanted to marry, I would first have to ask Charles Jewell and then see if he would build me a house. It was early in 1927 when I first asked Wood what he thought the chances were of my getting permission to marry from Jewell. He said it was worth a try and advised me to wait until Jewell came out on his annual visit rather than write to him.

In October 1927 I wrote to Kitty asking her if she would like me to come down to B.A. on November 11th for the Armistice Ball which was held in a large hall and run by the British Legion and, if so, could she find a partner for Frank. Of course she was delighted and said her great friend, Kittles Draper, who worked with her in the Otis Elevator Company, would partner Frank. How I longed for November 10th when Frank and I would set out by train for B.A. I did not say a word to Kitty about my coming down to ask her to become engaged to me.

At last the day came and we set off by car to San Jorge to catch the train to Rosario where we would catch the express to Retiro, B.A. We sat in the luxurious pullman until it was time for dinner. During the meal we shared a table with an elderly English gentleman. First, we had two rounds of cocktails and during dinner a bottle of Chianti and I was very excited, which amused Frank very much. In fact we were both very happy at the prospect of a Ball with only British people – so different from our life at the estancia where we never saw an English girl. The old gentleman broke into our conversation by warning us that the wine we were drinking was very potent and B.A. was a very wicked city for two young English lads. We told him that we were going to the Armistice Ball with two well-brought up English girls and that I was hoping to get engaged. He was very amused and treated us to a drink wishing us the best of luck.

When we arrived, we had difficulty in finding an hotel but in the end we found one which only had the bridal suite left. This consisted of a room with a large double bed and one very small dressing-room with a small bed. We both wanted the double bed, so we decided the best way to settle the argument was to go down to the bar and play dice, the best of three games. I won all three which was a good omen for me.

The next evening we met Kitty and Kittles, both looking very lovely in their long dresses. Not to be outdone, we were dressed in our dinner jackets, smelling of moth balls. When we arrived at the ball we discovered that it was a Programme Dance. I was surprised, as back in England your partner was expected to dance with you only. To prevent anybody 'cutting in', I wrote Kitty's name against every dance except two, when Kittles and I danced together. After the Supper Dance, the band played my favourite waltz, 'The Blue Danube' and while dancing with Kitty, I told her I had something to say to her of great importance, in private!

During the evening, I had noticed that the 'sitting out' arrangements were rather poor and as I certainly wanted to be alone with Kitty, I suggested to her that we should take a taxi and go for a ride to Palermo Park and the Japanese Rose Gardens. As we moved away from the dance floor, I gave Frank a knowing look as he knew I was going to ask Kitty to marry me. We waited outside until a taxi appeared. I think it was a Packard or a Buick, and told the driver we wanted an hour's drive. He gave me one look and immediately pushed up his front mirror. I thought that was very discreet of him.

When we arrived at the Japanese Rose Gardens, all floodlit and very romantic, I said, "Kitty, I have something to ask you."

Her reply was, "Jim, I know what you are going to say."

That gave me encouragement and I said, "Kitty, will you marry me?"

"Of course I will," she said. At that moment, to my horror, my nose started to bleed. It must have been a rush of blood to the head!

After an hour, we arrived back at the ball and Frank gave me a long look and the 'thumbs up' sign. I nodded and we sat down at an empty table where we were joined by the others and Frank ordered a large bottle of champagne. Then followed kisses and congratulations all round.

Kitty stayed in Kittles' flat and Frank and I returned to the bridal suite. I remember I did not sleep very well – only to be expected, I suppose.

The next day, Kitty and I went off to buy the engagement ring – a single diamond which, I think, cost 240 pesos. Today that amount would not buy a stamp!

On Sunday, Frank and I were invited to Kitty's parents' house. Her sister, Margaret and her husband, Kenneth Cordery, and the two unmarried boys, Steve and Jack were there. What a meal that was; one would think that someone had died! Mr Cork never allowed any frivolity during meals and you had to speak impeccably or you were pulled up at once. I must say I felt sad, having experienced engagement parties in my own home in England. Next day, Kitty and I visited the Youells and the Hansens, both families living in the suburbs of B.A. and what a difference in the atmosphere there.

* * *

In my job of supervising the Colonia Santa Anita, the most important part was to receive, in lieu of rent, a percentage of their total threshing of the crops. This payment in kind had to be the very best of each of the cereals that they had grown and threshed. My word, what games they would get up to, to avoid this obligation, filling half the sack either with earth or inferior cereals. It had to be the best or I would send them home.

Mr Jewell arrived out from England in early 1928 and I approached him concerning my prospects of marrying Kitty and of his building me a house. He replied that if the harvest was a good one in 1928, there might be a possibility of a new house. This, to me, sounded very encouraging. Little did I think he was just avoiding the subject for his own ends. How green I was to trust his spoken word.

In June of that year I was due for six months' holiday, with pay, in England. I naturally thought that I could marry Kitty and take her with me, but Mr and Mrs Cork advised me to wait another year to see if Mr Jewell would keep his word. So reluctantly, I had to sail off on the *Highland Laddie* from B.A., leaving poor Kitty to shed a few tears at my parting.

It was a delightful passage home with a good crowd of English people. There was plenty of sport during the day and dancing at night. What a change from Las Patacas to suddenly find oneself in such young company. In due course, we arrived in the Channel and what a thrill it was to see the white cliffs of Dover again. How strange that all the love affairs I have observed on passenger boats, seem to disappear like mist on a summer's day, when the coast of England appears on the skyline.

I had a wonderful reception from my family and Sir Arthur. My first lunch at Blagrove I well remember. It consisted of a specially reared duck that my father had fattened on barley and skimmed milk, green peas and new potatoes from the garden, followed by my favourite pudding, a trifle, and a good cheddar cheese.

My relationship with Sir Arthur took on a new form; no longer was I a little boy, but a grown man. I spent a few days with him at Youlbury and recaptured my old life with him. I knew that I could visit him at any time and if I was walking round the gardens or lake, I would pop in and spend a little time talking to him. He realized that my life was my own and that the family had first call on me. Gone were the days of Miss Wiggins and school holidays. What was nice to find was that nothing had changed at Youlbury; everything seemed to follow the usual pattern. Sir Arthur was still very interested in the Scout movement and the woods still echoed with the sounds of the voices of young boys, especially when they bathed in the lake.

Back in 1926, Sir Arthur, with the help of the Oxford Preservation Society, bought some ten acres of wood and heath on Boars Hill from a speculative builder who was turning that lovely area of Boars Hill into a housing estate. His idea was to build an earth mound similar to Silbury Hill and the general public would be able to climb up the stone steps and look down on the 'dreaming spires of Oxford'. This view from the mound was the first thing that struck you, but your eye could rove almost 180° which brought in the Chiltern Hills, the Thames Valley and the glorious Downs as far as the White Horse. Unfortunately, some time later, the mound slipped, due to the wrong soil being used. By the time I arrived back the mound, now called Jarn Mound, had been properly constructed and the stone steps and handrail were in place and on the top of the hill, on a flat concrete base, a small pillar had been constructed, in which a round copper plate had been inserted showing all the important roads, rivers, towns villages and historical sites, all based on the compass so that you knew exactly where you were looking.

The next step was to lay out the 'wild garden' to be filled later with wild flowers, plants and trees from all parts of England. On two occasions Sir Arthur and I, armed with boxes and trowels, set out by car, for various places to collect the specimens. All the paths were laid out by Sir Arthur himself, based on his delightful way of avoiding a straight line. He went ahead of his men, armed with short canes with which he marked out the routes they had to follow. Small ponds were constructed and a little fairy glen was included. On the side of the mound were bushes and blackberries and gorse (to keep children from clambering up the sides) and at the foot of the steps two lovely sweet briers were planted. Sir Arthur always plucked a leaf and pressed it to his nose as he clambered up the steep steps. I should mention that to help the elderly to climb up, Sir Arthur had a nice cast iron handrail built, but as a deterrent to small boys sliding down it, he had, at intervals, raised little notches, with wonderful results!

As Sir Arthur had gone off to Crete with Dr Mackenzie for further excavations at Knossos, I took the opportunity of visiting my sister Madge, now married to Foster Anderson and living in Riga, the capital of Latvia. Foster was manager of the United Baltic Shipping Company which was engaged in buying timber for pulp and pit props for the UK. He could speak Russian and German fluently. I took my mother with me, travelling by boat through the Kiel Canal to the port of Tallin, the capital of Estonia, and then by train to Riga, now alas, part of the Soviet Union. What an interesting and wonderful time I had, for, through Foster and Madge, I had the entrée to most of the legations. At the weekends we all went to a lovely wooden house among the pines on the Baltic coast. It was here that I first met Nina Fraudenfelt, a very great friend of Madge's. As Nina could speak four languages, I was put in her charge so that I could see the country and be entertained by her friends. I have never been to so many parties, dances, and nightclubs in my life. Here in Riga, I was a bit naughty for I never had time to write my long weekly letters to Kitty and instead I sent post cards. When I returned to Blagrove, I went back to the usual weekly letter.

On my return home, I took the opportunity of indulging in one of my greatest pleasures – shooting, not only at Blagrove, Youlbury and Upton, but also local farmers' shoots. It was a wonderful holiday.

Meanwhile, Sir Arthur had returned from Crete, so I spent a few days with him. It was then that he gave me one of his pictures, a delightful watercolour of Sunningwell village by A. J. Lambert, who had done so much work for Sir Arthur's massive book on his discoveries of the Minoan civilization, by drawing the small objects, such as rings and seals, etc. This picture, he thought, would remind me of happy days at Youlbury and my engagement to Kitty.

I sailed back to Argentina, 1st class again, on the *Highland Glen*. It was rather a dull voyage compared with the homeward one. I had told Kitty the time of my arrival, so you can imagine how excited I felt when the boat tied up at the dock and the passengers started to disembark. To my complete surprise and disappointment there was no sign of Kitty, but I did not give up hope until all the passengers had been met and had disappeared. There was I standing on the dock completely alone, feeling very foolish and still no sign of Kitty. I was really worried but then I caught sight of a face peeping round the corner of the customs shed and, sure enough, it was Kitty's! She had been there all the time, just watching me; in other words, paying me out for not writing letters to her and sending post cards only. I might add that I was quickly forgiven!

I stayed a few days in B.A. before returning to Las Petacas, just in time for the harvest which turned out a dreadful failure, so there was no hope of marriage till 1930.

In early 1930 James Wood, who had recently been married, had left for England on a 6 month holiday so Mr Jewell spent the next 6 months at Las Petacas, much to our disgust. He started a scheme which completely changed my life as it led to my eventual return to the UK for good.

The world crisis was beginning to be felt in the Argentine. Harvester combines had really taken hold in a big way and Jewell had been approached with a plan in which he would stand to win a considerable commission by selling combines to our colonists. After a large demonstration by four companies, I had to try and sell these machines to the colonists on the understanding that they had to pay in full at the end of the harvest. I knew for certain that some of the colonists were perfectly safe and about twenty of them signed contracts within a month.

When Wood returned from his holiday, Mr Jewell went back to the UK which was a relief to all of us.

One Sunday, I was playing polo with the staff and I had just got off my pony, whose name was Dolly, and was standing just in front of her face trying to keep the flies from molesting her eyes, when unfortunately, she suddenly brought her front hoof down with a stamp on my left foot. It happened that I was born with a hammer toe so you can imagine what pain it caused. I quickly removed my top boot and I could see that considerable damage had been done. The next day I went to San Jorge to our doctor, who advised me to have the toe removed in hospital in B.A. I was delighted with the idea as it would mean that Kitty would be able to see me every evening and also at weekends, at the British Hospital.

For some reason or other, my toe refused to heal properly. I think the trouble was caused by the use of iodine around the amputation. I was told at the hospital that I would be able to leave within a week, but it was, in fact, three weeks before I left. I stayed with the Hansens for a further week before returning to Las Petacas.

When I returned, I found that my assistant had sold six machines to colonists who I knew would never be in a position to pay for them. What made matters worse was that the value of their crops had fallen, due to the world recession. I told Wood what I thought would happen and he agreed with me but nothing could be done as the combines had already been delivered to the colonists. Meanwhile, the value of cereals continued to drop which made me very uneasy and I could see again my chances of marrying Kitty receding.

At last, the day came when all the colonists who had bought the combine harvesters had to repay the company, either in cash or from the sale of grain. Mr Jewell had already paid the manufacturers to enable him to gain the commission that they had offered. Sure enough, the colonists who had bought the machines during my stay in hospital, could not pay for them and they were returned to us, remaining on our hands for a long time. This state of affairs had to be reported to Mr Jewell and very soon a letter arrived from him to Wood telling him that I was entirely to blame for mishandling the sale of the harvesters and that I was to be given notice to leave. But, the letter continued, as I had given good service in the past, I could remain at Las Petacas for six months while I looked for another job.

You can imagine how I felt and I asked Wood if I could write to Mr Jewell and put my case to him, but he replied that it would be of no avail. Furthermore, the whole thing was one of Jewell's mad schemes and Wood, being away in England at the time, washed his hands of the whole matter. His advice to me was to look for another job so that I could marry Kitty. You can imagine how she felt having waited since 1927 and now in 1931, no prospect of our getting married. This was a dreadful blow to us and the depression was getting worse every day. I began to realize that all those promises that Jewell had made over the years, he had no intention of keeping.

I, at once, started to make enquiries in many directions and also asked our agents, Krabbe, King & Co., in B.A. to find me employment on other estancias, bearing in mind that I had some years experience in the grain trade.

The world depression continued and we heard that Mr Jewell, being a member of the Stock Exchange, had lost a considerable amount of money.

Meanwhile, some of our staff at Las Petacas had left and the time soon came round for me to pack my bags. One night, after supper, I decided to auction all my posessions for which I had no use, i.e. books, saddles, bridles, riding breeches, polo sticks, etc. Before starting the sale, I invested in a bottle of whisky Vat 69, which was soon consumed, with the result that all my belongings were sold at good prices. We then sat down and played our usual game of poker till nearly daylight and that, too, seemed to go my way. I always have happy memories of Vat 69!

I left Petacas towards the end of 1931, never to return, except in my dreams which were mainly of my unfair dismissal by Jewell over the combine harvesters. I did get a small commission from the firm of Massey-Harris! On arriving at Retiro Station, B.A., I was met by Kitty who was, naturally, very depressed by the situation I was now in.

For the next six months I stayed with the Hansens, hunting for a job in another estancia but all to no avail. The English companies all said the same thing; they had nothing for the moment but they would take note of my requirements. It was a dreadful period and I found other Englishmen and even major-domos all looking for work. To make matters worse, Mr Cork was putting pressure on poor Kitty to break her engagement with me. He pointed out to her that she had waited four years and still I was not in a position to marry and there were no prospects of my getting a job and that she would become a bad-tempered old spinster. To this Kitty replied it would make no difference and she was quite ready to go on hoping and waiting. Mr Cork then tackled me and said that as an English gentleman, I was bound to release Kitty from our engagement. This I refused to do and he told me that I would no longer be welcome at his house. Every night I would wait for Kitty outside the Otis Elevator Company, where she was the Manager's secretary. I can never thank the Hansens enough for all they did for me during the months when I hunted for a job.

Early in 1932, I was called to the offices of Messrs Krabbe King and Co. as they had a temporary job which might have good prospects in the future. It did not take long for me to present myself at their offices and I was told that the Condesa de Chateaubriand wished to interview me and would I go along to her palace in B.A. at 12 o'clock.

"Would this give me time to have lunch first?"

"Oh, yes," Mr King replied, "your interview is for 12 o'clock tonight." I made enquiries and found out that the Countess was a wealthy woman, divorced, with one child, a boy with a club foot.

On my arrival at the palace, which was built very much in the French style, I was conducted up the marble staircase by a liveried footman who told me to sit on a large sofa in the ante-room. Down one side of the cushions on the sofa, my hand encountered a man's pipe; the tobacco had been removed but the pipe was still warm! I had to wait ten minutes before the door opened and the Condesa came towards me. She was a very strange-looking woman; fat, and smothered with diamonds and emeralds, I noticed that she had a club foot, but most striking of all were her eyes, which were very large, dark and shining. She gave me to understand that she had a large estancia called Colonia Carmen which was rented to colonists under the same conditions that had existed at Las Petacas. Her manager and bookkeeper were Polish Jews, who had fled to the Argentine some years before, because of the persecution. Certain rumours, it seemed, had reached her and she felt that her major-domo, with the help of the bookkeeper, was robbing her and she wanted me to spend six months at her estancia in the province of Santa Fé to live with the family, to supervise the threshing of the maize harvest, keep my eyes and ears open and to talk to the colonists whenever possible. I was not to correspond with her and when the time came, she would ask her major-domo to send for me. At once, I could see that I had to be a spy and I asked her why she wanted me, an Englishman, and not an Argentinian or any other national. To my surprise, she replied that she had told Krabbe, King & Co., that she wanted an Englishman as he would be proof against bribery by her manager. A good salary was offered and my rail fare refunded. After thinking for a few moments, I decided that I would accept the job for, who knows, if the major-domo was thieving, I might stand a chance of getting the job which would mean that I could marry at last.

I must say, I was worried as I knew the Condesa had sent a wire to her major-domo telling him of my arrival and I began to wonder what sort of reception I would get for they must have guessed that I was being sent to spy on them. To my surprise, they seemed to welcome me and I was told that I was to live with them in their very nice estancia house. The ménage consisted of the major-domo, Valentine, his wife and two children, an unmarried sister, and the bookkeeper. Valentine took me round the colony in his car to inspect the threshing of maize and introduced me to the owners of the various machines.

My duties, for a considerable time, were to inspect the maize threshing and check the weighing balances to see that they were correct. I always made a point of talking to the colonists who seemed to be very intrigued that an Englishman was working for the Condesa. I soon became aware that all the colonists of every nationality were scared stiff of Valentine.

After about ten days, it happened that Valentine's wife mentioned at breakfast that she was short of eggs, butter, and fruit, and she also needed four young chicken as friends were coming to stay. Valentine asked me to come with him and threw two emtpy sacks into the car. I thought we were going into town but we went straight to the colony. When we arrived at one of the colonist's houses, it seemed to throw the household into a panic. They lived in very poor standard houses with mud walls and tin roofs. The colonist immediately swept off his hat, bowed to us and begged us to step down and partake of a glass of wine or maté. Valentine refused the drink, but, at that moment, some young cockerels came round the car looking for titbits. On seeing them, Valentine casually remarked that they looked very nice. At once, the colonist told him that he would indeed be honoured to present two of them as a gift. In no time, the children caught the two young cocks and they were popped into the bag. Then the same procedure was gone through at other colonists for two more chicken, eggs, butter and fruit! It became obvious that they were all scared of the major-domo and this seemed to me disgraceful, as Valentine was in receipt of a good salary and expenses.

It was not long before a more serious case arose. A certain colonist, according to Valentine, was not farming his land in accordance with his tenancy agreement and was a few months behind with his rent. A letter was sent to the poor colonist to say that he was to vacate his land within two weeks. The next day, the man turned up with his wife and four children and begged to be allowed to stay and be given time to find the money to cover the rent. I kept out of the way for the wife and the children were crying. Nothing would satisfy Valentine; the colonist had to go and, furthermore, a policeman was sent to see that they went off, taking their livestock and machinery with them. They would have to look for another farm to rent. I thought this would be a prima-facie case that the Condesa would be very interested in. Within a week, fresh colonists arrived at the office as they had heard that there was a vacant plot to rent, but they had to pay a thousand pesos to get it and that money went into Valentine's pocket. On making enquiries, I found that this sort of affair was fairly common especially when the leases expired and a new contract had to be negotiated plus a bribe!

By this time, I thought that I had enough evidence for the Condesa to throw the major-domo out, and I also heard that he had bought several houses in B.A. At the same time, I could not find any evidence of any malpractices by Valentine which would affect the Condesa, although I had my suspicions. I must say Valentine and his family treated me well and told me many stories concerning the Condesa and her addiction to gambling. Many cables would arrive from her from all over Europe for her major-domo to remit money.

At last, the day came when I received a message to return to B.A. I went to Krabbe King and was asked to present myself at the Condesa's palace at 1 a.m! At the time, I thought I was on a winner and told Kitty what I thought might happen.

I outlined to the Condesa the facts as I have already mentioned and she listened with keen interest.

"Mr Candy," she said, "it seems that my major-domo has not been robbing *me*, and if the colonists give him freely, the items you mention, more fool they; and if he can get perks for the letting of the land – good luck to him." I had to tell Kitty that, once again, I was back to square one.

I then took on a salesman's job, advertised in the papers, based on commission but this produced nothing and I was out of pocket at the end of the day. I don't know what I would have done without Mrs Hansen's kindness. I did manage to persuade her to accept a payment for board and lodging. During this period, I was feeling desperate and depressed and, to make matters worse, one day on Lomas station, I came face to face with Kitty's father and he cut me dead.

I had seriously considered taking a job as a steward on a boat to Canada, with Ernest Hansen who had been out of a job for six months. In fact, we had already been offered this job and I was about to tell Kitty, when out of the blue, a job was offered to me in a firm south of B.A. called La Plata Cereal Company of Bahia Blanca. My experiences at Las Petacas were of great help, but, being a 'new boy', I had to start at the very bottom.

On arriving at Bahia Blanca, I was told to report at the docks to where the goods trains from the interior, discharge their grain direct into the ships' holds. Oh dear! What dust there was when the hessian sacks were slit and the grain was poured straight into the grills and was then conveyed direct to the holds of the grain ships which were bound for European ports. I had to be at the port gates at 5.30 a.m., and waiting there, were people of every nationality under the sun, all hoping for employment. Any man lucky enough to get a job, would be given by the Capataz, (foreman) a numbered brass disc and at the end of the shift, he could cash this at the office. Of course, the number of jobs available depended on how many ships were in for loading, not many, I can tell you, as the depression was squeezing even tighter every day.

Each morning, there were about 200 men all lined up, awaiting the Capataz with great hopes of being handed one of the coveted discs. The first man in the queue was given one and the rest would be handed out at random; he would miss, say twenty and then hand out another one and so on until all the discs were gone. The unlucky men, very dejected, would return home to their families. It was very sad to see, but next morning they would be back again.

One day, we had to carry the sacks on our shoulders, run down the plank into the ship and drop them, in perfect fashion, to the man below who was stacking them. I noticed that an English company was installing sprinklers in the docks so I spoke to the Englishman who was directing his little gang and he was amazed to find a fellow countryman working under such conditions and said it was the first time he had run up against an Englishman working amongst all this dust, loading a ship. I told him I was out of a job and needed the work. Oh, was I not glad when the Company called me into head office and told me that I was to work out in the camp in a small town called Villalonga, to load wheat from the Russian colonists direct on to the trains! What a difference to be back in clean air again. With me, at this station, were other 'receivedors' from three large cereal companies and I became quite friendly with a receivedor called Francisco, a German, working for Drefus, an exporting company. Every evening, he and I went for a walk and for the first time I heard about Hitler. This was in 1933 just after Christmas. I used to listen to the German as we went for our usual walks from the town along a straight road with no turnings and he told me how Hitler was going to be the saviour of Germany. His great enthusiasm for Hitler impressed me very much and, at the time, I thought he was right. He quite agreed with the persecution of the Jews.

In January 1933, Mr and Mrs Cork returned to England, after thirty years in the service of the Great Southern Railway, leaving Kitty and her brother behind, living in a boarding-house.

In the Argentine, January is the month of Carnival, a three day holiday with dancing in the streets, decorated floats, ladies dressed in all sorts of costumes and masks, and everyone throwing streamers. There were bands, trumpets, water squirting, etc., all this going on till the early hours of the morning. I decided that I would catch a train to B.A. for the weekend so that I could see Kitty. While at Villalonga, I had grown a moustache, without saying anything to her. On my arrival at the station, Kitty was waiting for me but when she saw my new moustache, she marched me straight to the barbers and told them to shave it off. There was no way that Kitty would go about with a man with that thing on his face. I spent three happy carnival days with her in an hotel and then took the train to Villalonga to start loading wagons again.

* * *

I had been there about ten days, when I received a telegram, forwarded on by Kitty, from my sister, Madge, asking me if I would return back home to go into partnership with my father and brother, Martin, in a five hundred acre farm belonging to Lady Singer. That telegram changed my whole life, yet again. My first reaction was the realization that I could now, at last, marry Kitty and give up, for good, my life here in the Argentine. That night, I had a long talk with Kitty by telephone and I told her I would resign from the Company and marry her, as soon as possible, and return to England to take up farming. How excited she became that, at last, we could be married.

Within a few days, Kitty forwarded a letter from Madge, giving me more details of Lady Singer's farm at Milton Hill, near Didcot, describing the land, farm buildings, a model up-to-date dairy, a house for my mother and father, but best of all, a lovely little Norfolk reed thatched cottage with two bedrooms, for us two. It seemed almost impossible to believe that our luck had really changed.

I wrote to the La Plata Cereal Company, telling them that I would resign at once and that I was returning to England to farm. I packed my few things, said goodbye to my fellow workers and caught the train to B.A., where Kitty was waiting for me with joy written all over her face. I had previously sent a cable to my father telling him that I would get married and return at once. I asked Kitty how soon we could marry and book our passage home to England, and she said that as she had had the row with her father, it would be better to tell her parents of our changed prospects and ask their permission to marry. A cable was sent to them in Blackpool. Back came the answer, *Wait six months, letter in post. Dad*! In this letter was strongly worded advice that I should leave for England on my own, find out what my prospects were and then, at the end of six months, Kitty could come to England and marry me. Oh, how Victorian! Kitty then sent the following cable, *Can't wait six months. Marrying on 22nd April and returning with Jim.* How I laughed. "I bet your father will think you are already expecting."

More letters arrived from England, giving many more details concerning the new farm and what capital was available, etc. I had, at that time, saved over £1,000, which would be useful to invest in the new venture.

Kitty and I came to the firm conclusion that we would get married before we returned to the UK. We thought that to spend our honeymoon on a passenger ship would be a very attractive idea, for I knew that once we were at Blagrove, I would have to work very hard every day. We had to take over the new farm at Milton Hill on September 25th and hand back Blagrove Farm to the owners. Then we had second thoughts; why spend our savings on a passenger liner when we might, possibly, find some captain of a tramp steamer who would be prepared to take us? So we went to B.A. to hunt round the docks and after a couple of days, we found a captain who would let us sail with him. I think he agreed to our request when we explained our reasons and that we were getting married in B.A. before sailing. There was only one condition and that was we had to be part of the crew; I would be a steward and Kitty, a stewardess! In this way, we would fulfil the necessary regulations. He would charge £10 each, which, I expect, went into his pocket. We were to have the run of the ship, no duties whatsoever, all our meals would be taken with the officers and we could have a cabin to ourselves.

In the Argentine, no marriage was legal until you had been first to the civil marriage ceremony (registry office). Then you could go to the church of your choice but you had to show your vicar the certificate from the registry office. Only then, could you go ahead with a church marriage ceremony. On April 10th 1933, Kitty got leave from the Otis Elevator Company, but for the morning only and in less than 10 minutes we were married! I was very amused when my aunt came to see me and she begged that I would not sleep with Kitty until we had, in her eyes, been properly married in the Church of England at Temperley, where Kitty lived!

After seeing the vicar of Holy Trinity church, the day was fixed for the

22nd April, 1933. Charles Gummer was my best man and Nola Strong was Kitty's bridesmaid. There were not many guests, the ceremony was in the evening and reception was held in Kitty's sister's house. Soon after we left for England, Charles and Nola got married and we saw them in England on their honeymoon.

Our boat was due to leave three days after the wedding, but was delayed until April 30th, as she was loading grain for London. We spent the first two nights in the City Hotel but we decided to move into another hotel to save our money.

The day before sailing, we went down to the boat to take our luggage. We found out from the captain that the shipping owners were Quakers so there would be no drink on board. He told us that the cabin had only one berth, as it was intended for the pilot, should they need one. Well, it was a bit of a shock when the captain took us to our cabin for I could see that in no way could I get very near to Kitty so I asked him where I was to sleep and he replied, "On the sofa". This sofa was of the typical horsehair variety and I could see the first roll of the ship would shoot me on to the floor! The captain could see my point when I lamented that I had only been married for three days; could not something be fixed up so that I could, at least, sleep near to Kitty? We all had a good laugh and the captain sent for the mate and he, in his turn, sent for the bosun and finally, the ship's carpenter arrived and the situation was explained to him.

"Quite simple," he said, "I can fix up a perch level with the bunk with two horizontal supports." Boarding was fitted and the mattress was laid, with an added safety guard for me to be safe if the ship rolled. The cabin was now very small but we managed. The difficulty was for Kitty to scramble over my bed to reach hers and we had to have a small step ladder to get us to the right height.

At last, the day arrived when we set sail for the UK and we were seen off by the Hansens and a few friends. The voyage was pleasant but the ship was old and slow, but who cares under our circumstances? There was no refrigeration and when the ice box had all melted after the first week, we had to live on salt meat until we got to London. The crew were amused to see two passengers as this had not happened before. We passed the time reading, playing deck golf, shuffleboard and deck quoits with the captain and the officers. We travelled no faster than ten knots. The sea was calm; we had no storms. When we got north of Brazil, we had to stop for one day for repairs to the boilers. I have never seen such a calm sea, just like oil. There was not a breath of wind and it was boiling hot. It was very strange; the ship never moved and not a sound came from her, save the voices of the men as they lay about the hatches, as it was Sunday. The horizon seemed to stretch for ever and ever and there was not a ship in sight and no sea birds, just a vast stretch of water. We were very glad to hear once more, the throb of the Queensbury engines. Next day, we picked up the Trade Winds that carried us, in the end, to Hays Wharf, a very old and run down dock in London. On looking down from the bridge, we were astonished to see a Rolls Royce and a chauffeur standing beside it. This intrigued all of us. What was a Rolls Royce doing in such a decrepit place? When we were tied up and the gangway lowered, a lady stepped forward and, lifted up her head, said, "Jim, is that you?" My God, it was my sister, Madge, come to welcome us. She wanted to take us off the ship at once to her home in the West End, but the captain refused to let us disembark as we had to sign off in the Seaman's Mission with the rest of the crew next day.

When we had picked up the pilot off Gravesend, he had with him, a telegram for us from Mr and Mrs Cork saying, *A loving welcome awaits you.*

What a lovely feeling to be back in London in early June with all the flowers in full bloom in the parks.

After a couple of days, Madge, Kitty and I set off, by train, to Blagrove. Martin met us at Culham station in his car. After the usual greetings, he told me that father had reared some Aylesbury ducklings and that we were having roast duck for lunch, green peas and new potatoes from the garden. My mother was quite overcome when we stepped out of the car and she saw Kitty for the first time.

The following days were a great trial for Kitty as troops of relations and friends came to have a look at my bride from the Argentine. At the time, I did not realize how she felt. She had never before seen all these people in her life and she felt very shy. In fact, she felt like running away and it was only when my Uncle Will came that she felt here was a person that she knew well in B.A. The shyness gradually went and when she saw our sweet little Norfolk reed thatched house at Milton Hill, with all mod cons and a nice garden with fruit trees, etc., and where we could be alone, she became much happier.

Soon after I had settled in at Blagrove, Sir Arthur returned from excavating in Crete. I had written to him before leaving the Argentine, telling him of my marriage and return to England, and it was not long before he was on the telephone inviting Kitty and me to luncheon with him. Knowing the etiquette at Youlbury, I made Kitty wear a hat for lunch. Sir Arthur was delighted to see me and wished me every success in my new venture. In his usual way, he conducted us both round the gardens and lake. We had a very nice wedding present from him, a cheque for £100.

On October 1st, we took over the tenancy of Lady Singer's farm at Milton Hill. Home Farm was indeed a godsend after Blagrove. All the farm buildings, model dairy, the three cottages and a very nice office, were the envy of our farmer neighbours. We had an arrangement that Martin would be responsible for the working of the arable land and the poultry; I would keep the books and pay the wages and, above all, control the milking and distribution of milk in Abingdon. My father would be semi-retired, but he would build up the hay and corn ricks. He had started milk retailing in Abingdon in 1906 with a horse and cart, a seventeen gallon churn and a hand-serving bucket with two

Cloford Manor, Somerset – one-time home of the Candys

The author aged two at Blagrove

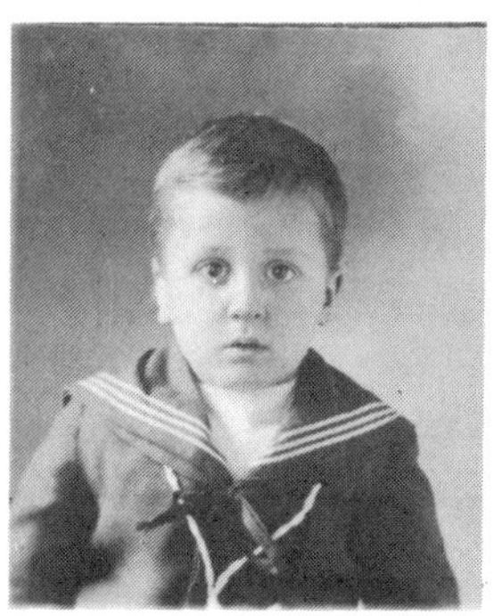

. . . and aged three

. . . and with Jock in 1913

Blagrove Farm

The author with his brother Dick at Blagrove

Author's mother and father

Sir Arthur Evans

Picnicking at Cadir Idris with Sir Arthur Evans

Youlbury, Sir Arthur Evans's house

Entrance hall at Youlbury, with the Minotaur set in the floor

1914-18 War Memorial at Youlbury

Duncan Mackenzie, a frequent visitor at Youlbury

Sir Athur Evans (right) with one of his many visitors at Youlbury

Land Girls on a milk round during the First World War

Estancia La Isabel, 1922

Branding cattle

One of the Show bulls at La Isabel

Branding cattle

Videla and Bonifacio, La Isabel, 1922

Mail and supply coach, Las Petacas, 1924

The author at Las Petacas

The author at Las Petacas, 1926

Gauchos at Las Petacas

Gauchos in their distinctive dress (above and below)

The author becomes engaged to Kitty in 1927 – who is seen below at Mar del Plata

Wedding Day – 1933

Our first home at Milton Hill Farm, 1933

Happy days, June 1937

Dairy, Home Farm, Milton Hill – 1933-6

Northcourt Farm
tithe barn, 1280

Northcourt Farm
Abingdon, 1950

Mayor of Abingdon, 1962/3

With Kitty at the Mayor's Ball, 1962

The author with some of his grandchildren on his 80th birthday

Memorial tablet to Sir Arthur, Jarn Mound, 1978

measures, half a pint and one pint!

I soon realized that we were very under capitalized, even with my money and that which my sister, Madge, had advanced us. Furthermore, the world-wide depression showed no improvement. Corn prices were at rock bottom; hay, which we had in abundance, was only worth £1.10s. per ton!

On June the 6th 1934, Kitty had her first baby which we christened Michael Arthur, after Sir Arthur. I asked him to be godfather, but he declined, due to his age, he said.

The financial side of our farming was very worrying, to say the least, as the price of cereals continued to fall. I had to curtail the amount of money that I paid out to my brother and my parents. We all had to manage on £1.10s. a week, but we were allowed extras in the way of milk, eggs, poultry, rabbits, hares, and other game. The milk retail side of our farming did manage to show a profit. Kitty was an excellent cook and housekeeper and she even managed to save five shillings a week out of the amount I gave her, for a little maid to take Michael out for a walk each day.

Nearly every week, Sir Arthur used to telephone me with these delightful words, "Jimmie, er, er, er, would you like to sup with me?" He liked to be reminded of the past we had spent together and to play billiards. At Christmas he always sent us a Christmas hamper from Fortnum and Mason. How the children looked forward to this wonderful treat. If I was ever ill and in bed, he always sent his chauffeur over with two dozen oysters. How I enjoyed those weekly evenings at Youlbury, for Mrs Judd always gave us oysters followed by game, in season, champagne, my favourite pudding, followed by port and glorious coffee!

In 1936 I had a touch of flu and Sir Arthur motored over to Milton Hill to tell me that I must have a holiday and recommended that I go to Lucerne for ten days, and that he would pay all expenses including the flight from Croydon to Le Bourget. And so I went. In those days, planes had only two engines; I must have fallen asleep, for, on waking, I looked out of the window and, to my horror, saw that the propeller had stopped. I thought that there was something wrong and that the pilot might not know so I called the air hostess and pointed it out to her and told her to tell the pilot what had happened. All she said to me was, "Look down there, Sir," which I did and there was the runway approaching fast and we were about to land. The propeller that I had seen was just 'feathering', the usual procedure before turning into the runway. What a fool I felt, but it was the first time that I had ever flown.

Meanwhile, farming at Milton Hill had not improved. We had an overall loss in all branches with the exception of milk distribution. My brother, Martin, had also married and he and Molly wanted to start up on their own and take up poultry farming. This they did, taking with them all the poultry and appliances that Martin had built. Then came a bombshell, when Lady Singer's butler came over to my father's house with a message that she would like to see him. My mother and father were soon round to see me. I could see that they were very disturbed and after he told me what had happened, we both came to the same conclusion. To me it was obvious that it would be concerning the rent. At that time, due to the bad state of agriculture, we were two years in arrears with the rent! My father asked me to go and see Lady Singer and to say he was not well! On arriving at the Manor, the butler led me into the large library and after a few minutes, Lady Singer came in. She said how sorry she was that my father was not well and how pleased she was when she met him about the farm and listened to his adventures in the Argentine, all those years ago.

Before meeting her I had discussed with my father what steps could be taken to reduce the arrears in rent. We could sell Bluebell and Daisy, two heifers and two ricks of hay, which would leave one year in arrears.

Lady Singer made no reference to the rent. To my complete surprise, she told me that her intention was to sell the farm, her manor, eight cottages, the riding school and five hundred acres of land. I nearly fell off the sofa! It seemed she had a small house in Torquay and wanted to live there for the rest of her life. The asking price was £40,000 lock, stock and barrel. She also said she had a prospective buyer at a higher price, but as we were her tenants and she liked my father, we could have the place at £40,000! After I had gulped and recovered my surprise, I said I would consult my father and would let her know our answer. To this she agreed and told me there was no hurry.

On my return to my father, his first words were, "Does she want the rent?"

"No," I replied, "she is offering the farm and all the buildings for £40,000."

"Good God!" my father exclaimed. "We can't even raise 40,000 farthings!"

After a week, I returned to Lady Singer and told her that, after careful consideration, we would have to refuse her kind offer. Things took their normal course, with a letter from her agent giving us notice to quit on the 25th September, 1937.

* * *

My father now felt it was time to retire completely from farming so he decided to have a sale at Milton Hill. As soon as I received notice, I set about finding a small dairy farm, in order to hold our present milk distribution in Abingdon.

In November 1937, our second son, David Yeoman was born. By Jove, he was a whopper; just over 11 lbs at birth! There were no problems with him; he was what we call in the dairy industry, a 'doer'! Kitty had no trouble feeding him from the very start. I remember when he was in his high chair at meal times, we had to put him between us so that we could both spoon the food into him. He reminded me very much of a young cuckoo always with his mouth open ready for food. As we had two small children, Woggie Madge gave us the pram which had been used for the twins.

When David could run about, he followed me one morning into the cart shed where I kept my bullnose Morris car. I lifted him into the passenger seat and started up the engine in preparation for backing out of the shed and turning sharp to the right so that with the next move, I could proceed out into the road. What I did not know was that the passenger door was not closed and that David was standing on the seat. I started to reverse and, at the same time looking over my right shoulder to see how far I could go before moving forward. Suddenly, I heard a scream from David and, on looking to my left, I saw that the door was open and no sign of David. I jammed on my brakes, dashed round to my nearside and, to my horror, I could see that the front tyre had skidded off his forehead – had it been another two inches, it would have crushed his head. I grabbed him, but I had not noticed that some of his front hair was under the tyre and in snatching him up, left a bunch of hair under the wheel. What a close shave that was! Kitty kept the hair for many years afterwards.

In 1937, I heard that Northcourt Farm, in the Borough of Abingdon, was to be let. The small farm was situated on an island of about five acres of grass and farm buildings belonging to Mrs Tatham. This was just what I wanted and there was another area of land adjoining, of about fifty acres of arable and grass to let. There was a tithe barn, built by the monks of Abingdon in 1280, a cowshed and barns of Berkshire type which were built in 1750, some large stones of which were believed to have come from the old Abbey in Abingdon. All the buildings, including the old bothy were listed as Grade II under a Preservation Order.

The only snag was that there was no farmhouse but there were vacant plots of land for sale quite close to Northcourt Farm, and I bought a plot of half an acre for which I paid £150! I approached Lady Singer and asked her if I could continue to live in my cottage at Milton Hill until my house was built. She was very kind and let me live there, free of rent, and in March 1937, we moved into our new four bedroomed house.

Before the sale of our stock at Milton Hill, I was allowed to take out and keep at Northcourt, six of our own breeding milking cows in full production, four in calf heifers, six yearling heifers, some young calves, a short-horn bull and one horse, a wagon, a tractor and various other farm implements. This was just enough to maintain my milk output of thirty-six gallons which I sold in Abingdon and a few outlying villages. The actual cash I had in Barclays Bank when I started on my own, was just £50! Everything else at Milton Hill was sold to pay off our debts and the arrears of rent to Lady Singer. Even then, there remained my father's debts of £300 owing to friends which, after several years, I was able to repay, much to their surprise.

Now followed a period of several years in which Kitty and I had to work all the hours that we could – not even a Sunday off, as the cows had to be milked twice a day and there was stiff opposition in the milk distribution business from six other milk retailers in Abingdon. On Sunday evenings, I had two hours of making up customers' books. The garden had to be looked after so I had very little time to play with Michael and David. My sister Madge and I had also to help mother and father who now lived at Harwell, paying the rent and rates and a married couple who looked after them.

I had also decided to have an attested herd of cattle in order to supply my customers with tuberculin-tested milk. Professor Huxley had started supplying TT milk in my area and I had to keep up with him! My heart ached when the results were known as I had some 'reactors' which I had to sell at once. Sir Arthur came to my help by giving me a cheque, thus enabling me to buy from an attested cattle sale. I remember the awful day when my best milking cow fell into a ditch and died, at the time she was producing five gallons a day. Here again, Sir Arthur came to my rescue.

As time went by, I began to build up my customers, due to the safety and quality of my milk, but this, in its turn, meant that I could in no way, produce the extra gallonage from my own herd so I had to buy in from another TT herd. I was fortunate in that I could acquire some from the herd of J. Benson who had a farm within two miles of Northcourt.

In the autumn of 1938 Kitty was pregnant again and we were quite sure that this time it would be a girl. In June, 1939, I took her to the nursing home on Heddington Hill, Oxford. Dr John Fisher, who had attended her for Michael and David, was with her this time. His wife had just had her third boy and they, too, had wanted a girl. Kitty had her baby and only a few minutes after the birth, when she had hardly recovered, she heard Dr Fisher exclaim to the midwife, "Oh dear, Mrs Candy has another boy." Kitty, at once, opened her eyes and said, "Never mind, doctor, I shall have one more try."

When Kitty was ready to leave hospital with Philip, David went down with whooping cough. This meant that I had to send Kitty to my mother and father. I was fortunate to have working in my office, a friend of my family, Enid Slade, who promptly stepped in, taking over the children, running the house and, at the same time, keeping the office going.

I remember so well having Philip christened in St. Helen's Church for when we came out of the church, we saw posters proclaiming that Russia and Germany had signed a non-aggression pact. At the time, I told Kitty this was a very grave state of affairs affecting, not only the Western nations, but the whole world.

When war was declared, almost at once, I had two of my roundsmen called up; in fact, one did not complete his milk round and I had to step in and finish the job. I had to take over for some time which meant that I was under considerable pressure producing extra milk because of the influx of evacuee children from London.

Then came the evacuation of the remnants of our army from Dunkirk and the call from Anthony Eden to form Local Defence Volunteers, later called the Home Guard. I immediately went to the police station in Abingdon and enrolled. I had in my posession my brother Dick's .45 colt revolver from the 1914-18 war. You have no idea how well I was received by the police as they had one box of ammunition which they gave me and, what was more, I was allowed to carry my revolver when I was on parade. Within a week, the whole of the force for Abingdon was supplied with six denim overalls, six forage caps and six armlets with the initials L.D. on them. With this went six rifles of the 1914-18 period and six rounds of ammunition! Each squad of six men had two hours of guarding Abingdon bridges and the perimeter of the RAF Station, on the look-out for German paratroopers, who were expected to land on the runways! After the two hour stint, the six of us marched back to the police station and handed our denims, rifles and our six rounds of ammunition over to the next squad. One terrible day, one silly ass, unloading his one round of ammunition, accidentally pulled the trigger and hit the ceiling. This now meant that the whole force of fifty men had only five rounds left between the lot of them! I was lucky as I had my revolver from which I would not part. Out of the six denims, I was invariably given the one obviously made for a sixteen stone man, which I had to wrap round me and which Kitty had to fix with a safety pin, and my forage cap was so small that if I sneezed or made any sudden movement, it would fall off. Kitty and the children would come down to the police station and peer round the corner and be convulsed with laughter as they watched me march off to do my patrol.

In time, things improved with more rifles, ammunition and uniforms. I was asked to retire as I could not attend the drills, etc., as I was farming, producing and retailing milk. I then offered my services as a Fire Watcher in the Northcourt area.

I was able, thank goodness, to employ girls from the Women's Land Army, and without their help I could not have carried on. They did a wonderful job during the whole of the war, milking cows, hay-making and harvesting, etc. Hats off to the Women's Land Army!

I must, at this point, mention two of my Land Army girls, Marjory Horden and Vera Raymond. Neither of them had any previous experience of working on a farm in any way. Marjory was terrified of cows, and yet she became a good milker and even helped with the calving. She became a friend of all the family, looking after the children when we were able to take a holiday. Now, she is a grandmother with five grandchildren.

Vera worked on the farm, hay-making and harvesting. She became an excellent roundsgirl and was particularly good in building up my retail milk trade and she was responsible for a small retail shop. Later, after returning from living in Australia, she came back to work for me and was employed in the accounts department until she left to work in London.

Here, in Abingdon, we were very lucky during the war. Only one German aeroplane dropped a stick of bombs which did little damage as the bombs fell on open land; all but one, that is, and that fell not half a mile from the farm, damaging a house. The occupant of the house was wounded as he was having a bath.

I continued my visits to Sir Arthur who still lived all alone in that large house, being looked after by Emma and Mrs Judd. I was with him the evening we heard that Crete had fallen to the Germans and I asked him if he was not afraid that the Germans would steal or destroy the treasures from Knossos.

"No," he said, "the Germans have a department in their army which is responsible for preserving, where possible, archeological sites. This also applies to our army, this department being under Sir Mortimer Wheeler." He said that the Cretans would carry on the war in Crete from the mountains, as they had always done against any invader.

In 1941, I noticed that Sir Arthur was becoming very frail, but he was as clear in his mind as he had ever been. Slowly, he began to sink and he had difficulty in swallowing food. Emma used to telephone me to bring him some cream so that she could make real ice-cream, of which he was very fond. I had to be very careful, for making cream was forbidden. I saw him for the last time, lying in bed, three days before he died. He made one last supreme effort by getting out of his bed and dressing in his ordinary clothes to receive a deputation of learned societies, from whom he received an illuminated manuscript. He, in turn, gave them a detailed map of Roman roads in and around Oxford.

On July 8th, 1941, Sir Arthur, my ever to be remembered friend, died.

The war dragged on. Rationing became very severe but, as a farmer, I was far better off for food as I was able to have eggs, milk, rabbits, pheasants and one pig a year plus extra petrol for selling milk round the district.

I was delighted when Kitty told me that she was pregnant again. At once, I went down to the Billeting Officer and asked him to remove a Civil Servant, evacuated from London, who had been billeted on us, which he did. On March 23rd, 1943, our daughter, Jennifer, was born. How glad Kitty and I were that, at last, we had a little girl. I remember that evening, dashing off to the hospital in Abingdon, on Kitty's old bike, to see the new baby. When I arrived, I left my bike with the others and hurried up to the ward. When I came out of the hospital, I took the first bike with a basket carrier in front that I could see and set off for home in great excitement that we had a baby girl, at last. Next morning, I was taken aback when I was confronted by a policeman who had come to see me concerning a missing bicycle! I went round to the garage and found, to my horror, a brand new bike and not Kitty's old one. It seems that another husband had been to the hospital that night and, to his annoyance, he found that his bike had gone. After making enquiries at the hospital and finding that all the visitors had left, he had been forced to take Kitty's old bike, in order to get home. I was able to satisfy the policeman that what had happened had been a mistake. He gave me the name and address of the owner and I set off to return the bike and picked up

Kitty's old one and that was the end of the matter.

1943 brought to us the first personal tragedy of the war. My sister, Madge, lost her only son, Donald Foster Anderson. He was killed in Tunisia after stepping on a mine, he was only 22 years old.

* * *

At last, the war finished with the surrender of the German armies in 1945.

In 1946, Madge rented from Dame Eva Turner, a beautiful villa on Lake Lugano, for two months, and invited Kitty and me to spend a holiday there with her. This was the first holiday we had had since 1939 and what a holiday that was. Switzerland was 'a land of milk and honey'.

With Madge, at the villa, was her friend, Hella Kerte, a singer from Richard Tauber's company. Hella was an inveterate gambler and, one night, she asked me to accompany her to Campioni, a small enclave of Italy, which consisted of a landing stage, a small village and a casino. We engaged a boatman from Brusino to take us down the lake at 9 o'clock and to return us home at 12 o'clock. Hella and I agreed to take only £5 each with us. By 10 o'clock, we were broke and there, on the landing stage, we had to sit until our boatman came at midnight. We did have a couple of francs for coffee, but the rest of the time, we just talked. We were discussing gambling and Hella told me that she could not break the habit. She then revealed a startling piece of information. It seems that while with Richard Tauber's company, they spent a season in Monte Carlo. Every night, after the concert, Hella would take herself off to the casino to play roulette. She became very interested in one particular woman who spent large sums of money every night. Generally, having lost her little bit of money, Hella used to sit beside this woman and, in time, they became friendly and, on occasions, the woman would win large sums of money and would give Hella a handful of chips to play with. My curiosity was aroused and I asked Hella to describe the woman to me, which she did. She was fat and had one leg shorter than the other and wore diamonds round her neck.

"What was her name?" I asked.

"The Condesa de Chateaubriand," replied Hella.

Now I knew why Valentine of the Estancia Carmen, where I had worked all those years ago, was always being called upon to remit money to the South of France.

On Sir Arthur's death, I inherited one tenth of his estate and two very nice water-colours by A. J. Lambert. Sir Arthur's affairs were in a very bad state when he died, due, in part, to his excavating the Minoan site at Knossos at his own expense. Today, museums jointly undertake the excavations of archaeological sites. I remember Sir Arthur, towards the end of his life, remarking to me, one evening after dinner, that he should now be living at his lodge gate and not at Youlbury. After some considerable time, I did receive from his estate about £1,500. With this money I was able to pay off the mortgage on my house that I had built in 1937. The mortgage was over twenty-one years at £7.10s per month!

After years of hard work, I had now improved my position in the milk distribution business, spreading out to other villages. At the same time, I realized that it was important that I should have security of tenure on Northcourt Farm, so I went to see Mrs Tatham to ask if she would sell me the farm and the five acres of grassland. This she was happy to do and offered me the tithe barn and all the farm buildings for a total of £1,600. I accepted without a word, for I could see the potential in years to come. This was realized in a very few years, thus 1946 was an important year in my life. Now, I could even take a day off to shoot with my friends. I shall never forget the wonderful partridge shooting at Chilton with Jack Waugh, and the delightful rough shooting with Bill Drysdale at Park End Farm at Radley, and neighbouring farmers; also the fun we had on the sweep for the total bag of game. What happy days with my shooting pals.

It was about this time that the great housing programmes began to get under way which affected me, inasmuch as the land round Northcourt was in great demand for building. I only rented this land, without any contract with the owners, and I was just asked by them to give up farming and remove my cattle. In time, this would have left me with only the four acres of grass that went with the dairy, so I decided to sell my attested herd and concentrate on milk distribution only. I approached the Milk Marketing Board to allocate me certain milk producers to satisfy my requirements. I greatly benefited from this housing boom; my milk sales increased as more people moved into my area. Naturally, there was great competition for this trade from other milk distributors in Abingdon, especially from the Argyle Dairy which was owned by William and Arthur Smith who also had interests in Wantage and Witney. It became a great challenge to catch the new owners as they took posession of their houses. We used to remove each other's bottles and hide them in the bushes with the hope that the customers, on opening their back doors, would see the right milk and card!

When the compulsory pasteurizing order came in, I decided to instal my own plant and rotary bottle fillers, etc. For this I was able to raise the capital by selling the tithe barn for conversion into a church. Previous to this, I was asked, mainly by demobbed soldiers, to sell part of my field for a football club which I did. The next of my fields was sold to the local cricket club. These three sales netted me about £5,200.

I was fortunate to have in my employ an excellent man, Harry Thomas, who, soon after the war, married one of my land girls, Jean Randall. He was in complete charge of all the plant and machinery and was quick to learn my 'batch monel' pasteurizing plant process. I was also lucky to have a very good worker in Johnny Latch, a Yugoslav prisoner of war, lent to me during a large part of the war. I was able to help him become a naturalized British subject and he married one of my girls from the dairy.

In the early fifties, we began to run into a serious shortage of man power, due to the motor industry. Even when we secured men from the North and provided accommodation for them, after a few weeks, they would leave and join the MG car works in Abingdon or at Oxford.

On one particular occasion, Kitty and I took a holiday in Cornwall, leaving our son, David, in charge. We had only been away two days when he telephoned to say that two of our roundsmen had left without a week's notice. That night, we set off home by car and arrived back at 3 a.m. To keep things going, I took one milk round and Kitty did the other. This she did for six months without a break until I could find someone to replace her.

One Sunday, Philip, who was a day boy at Abingdon School, went with Kitty to help deliver the milk and collect the money. This particular Sunday, Kitty was driving one of the large electric milk trucks and she had parked it on a slight slope. On getting out of the truck to reach for some milk in the crates, she had put her foot close to the front tyre. Evidently, the hand brake was not fully on and the truck ran back and pinned her foot beneath the large tyre and she could not withdraw it. Fortunately, Philip returned to hand over the money he had collected, saw his mother's predicament, jumped up into the cab, released the brake and drove forward releasing Kitty's foot. At that time, he was fourteen years old and had some knowledge of how the electric truck should be driven. Kitty carried on with the delivery and it was only when she returned home and took off her shoe, she found it was full of dried blood which had been squeezed out from under her toe nails. Next day, she carried on as usual and it was six months before we could find and teach another man to do her work. This showed me what Kitty was made of and she kept the home going at the same time. The reward from me for what she had endured, was a solid gold bangle.

My eldest son, Michael, had decided that milk rounds were not for him, so after leaving Abingdon School, he decided to go to Canada which he did in 1952. He stayed with my brother, Dick and his wife, Claude, until he got a job with the Bank of Nova Scotia and there he has remained ever since. It was a wise choice, for he has done very well indeed. He is happily married to Maureen and has two children, Paul and Lisa.

David also decided that he was not interested in the dairy trade and, for a time, worked in the car industry, delivering cars. This was getting him nowhere and eventually he decided that Kenya might suit him. So in 1956 we fitted him out like his brother and paid his passage to Nairobi and he, too, soon found a job on the tea plantations. He, like Michael, has done well and is now an expert in growing and processing tea and is Manager of a large tea estate near Lake Victoria. He is married to Mhorag and has two children, Natasha and James. I am very lucky to have three delightful daughters-in-law and a very good son-in-law.

I was now retailing five hundred gallons of milk a day and bottling for two other dairymen, making a grand total of six hundred gallons a day. Argyle Dairies and myself were the main retailers in Abingdon – all the little men had been bought up either by me or by the Smith Brothers. It was becoming absurd that Smith Brothers and myself should be in competition, running over the same streets and it was not long before we decided that we should amalgamate. This we did in 1956, and under the Smiths' manager, Fred Woodley, we became a very useful team. We were able to build up a very efficient business, for into the pool the Smiths brought in their retail branches at Wantage and Witney. During all the years that we worked together with Teddy Ballard, our Chairman, there was perfect harmony.

Having Fred Woodley relieved the pressure on me and I was able to undertake other interests, such as the local Dairyman's Association, of which I became President for one year. I was also invited to become a member of the Abingdon Rotary Club, and the District of Abingdon Chamber of Trade.

When I was President of the Chamber of Trade, at the Annual Dinner, the chief speaker was Air Marshal Macneece-Foster and he told me a lovely story of when he was Officer Commanding RAF Station, Abingdon, during the early part of the war. King George VI and the two Princesses, Elizabeth and Margaret, were invited to lunch at the Officers' Mess. Food was rationed and the King made quite sure that his family had the same rations as everyone else. Now the officers decided that they would forego their butter ration so that the royal party could help themselves to the butter on the high table. The seating was so arranged that the CO was on the King's right and Princess Elizabeth was on the right of the CO and naturally all the conversation was addressed to the King. During the meal the CO suddenly felt a nudge on his arm and, turning to the Princess Elizabeth, he asked her what she wanted; whereupon, the Princess asked him, in a whisper, to pass her the butter. "You see," she said, "we don't get much of that at home."

It was not long before Ernest Nicholson, the Abingdon Town Clerk, came to me to ask if I would stand as a Borough Councillor for Northcourt Ward. At first I refused, as I thought it might conflict with my fellow Directors, for political party feelings were very strong among the members of the Borough Council, and it might have a detrimental effect on business. The Town Clerk came back to me again and tried his hardest to persuade me to represent the people of my Ward. I did not like the idea of knocking on people's doors and kissing babies, etc! In the end, he produced his final plea by saying that I would not be opposed by anybody. Finally, I agreed and I was elected unopposed in 1952 as an Independent.

Kitty and I decided that Jenny, like the boys, should be given the chance to go abroad. After leaving St. Margaret's School in 1960, we were able to get her into a bank in Abingdon and then, when she was eighteen, we paid her passage to Canada where she stayed with Dick and Claude until she found a job in a Canadian bank in Vancouver.

On Jenny's first day in the bank, she, unfortunately, pressed the alarm bell close to her till. Within seconds, two Canadian Mounted Police dashed into the bank with drawn revolvers, looking for the bandits. Poor Jenny, what a red face she had!

In 1955, I had completed my three year period as Councillor for the Northcourt Ward which meant that I could put up for re-election for another three years. As I had represented my Ward unopposed, I was anxious to find out if my constituents were satisfied with what I had done during the three years. I put my name forward but, this time, I was opposed by Labour. This meant that I had to work hard to retain my Independent seat. This consisted of knocking on every door and chatting to the voters and leaving them with a copy of my policy on what I thought would be of interest, not only to my Ward, but Abingdon as a whole. I was well supported by friends, especially Mrs Cox and her son, Ron, who lived in the Northcourt Ward. Mrs Cox had recently been Mayor, the second woman ever to be elected as Mayor in the history of Abingdon. It was great fun dashing round my Ward getting people out to vote, checking the polling booths to see who had been in to register their votes, checking our lists of those whom we thought were my supporters and even offering to fetch them by car and return them home afterwards. Kitty and I were at the voting booth to watch the count. How delighted we were when I won by a handsome majority.

Meanwhile, the business flourished so we were able to take holidays. We very much enjoyed our visits overseas; rewards, after working so hard for many years. We had lovely trips to my sister Madge on Lake Lugano. Also to Canada, Kenya, New Zealand, and many countries in Europe.

Another three years went by and I became an Alderman and so was freed from any further elections and I had only to wait until my turn came round to become Deputy Mayor.

In 1956, Queen Elizabeth came to Abingdon to open the County Hall, after the work of restoring and repairing had been completed. The hall was built in 1682 by one of Christopher Wren's master pupils, a man named Kempster.

It was a wonderful day for Abingdon and the town was thronged with people. All the members of the Borough Council and their wives were introduced to the Queen, by the Mayor, Alderman Stow, in the Roysse Room. This was followed by a lunch in the beautiful Council Chamber, the walls of which are hung with lovely pictures. The lunch consisted of melon, served on a fine collection of Elizabethan platters, followed by roast beef (Aberdeen steer) and then followed fruit salad and cream, cheese and biscuits and, finally, coffee.

The Mayor had asked me to supply the cream which must be nice and thick, from Jersey cattle. The Queen and the Mayor sat at a small table beneath the Gainsborough portrait of George III and Queen Charlotte. Kitty and I were fortunate to be sitting quite close to the Queen and the Mayor. Naturally, we were most interested when the fruit salad and cream were served. The cream, which was in a George II silver jug from the Corporate Plate, was placed on the table in front of the Queen. Our moment came when the Mayor lifted the jug and offered to serve her. To our great disappointment, she lifted her right hand and waved her forefinger at the Mayor and said something that I could not hear. When I got the chance, I asked him what the Queen had said. Her reply had been, "Oh, Mr Mayor, I have to think of my figure." Bang went my chance to say to all and sundry who bought my cream, "As served to her Majesty the Queen."

Having been on the Abingdon Borough Council for ten years, three years as an Alderman, one might feel that, in time, one could be given the honour of being made the Mayor of Abingdon. In 1952, George Burrett and myself were elected as Councillors and then Aldermen, so, in 1961, when the Deputy Mayor had to be elected, we were both eligible. Whose turn would it be, George's or mine? The Town Clerk solved the problem by choosing us in alphabetical order; hance George became the Deputy Mayor. When, in the following year, George became Mayor, he asked me to be his Deputy Mayor.

Now, the duties of a Deputy included, not only understudying the Mayor, but also undertaking any assignments that the Mayor was unable to carry out, due to the tremendous number of functions he had to attend, and, also, to step in on full Council nights to take the Chair, if the Mayor was away. In other words, to be a general dog's body!

On one occasion, George Burrett asked me to represent him at the annual bowling match for the Preston Cup, between Abingdon and Oxford, as he was going to a Garden Party at Buckingham Palace, and he asked me to wear the Deputy Mayor's silver chain. The President of the club was there to meet me and he took me up the slight bank above the bowling green to watch the match. Just behind me was a low privet hedge and it was not long before I was aware of talking behind me. I turned round and there were two boys standing behind the hedge. One was quite a tall boy, about twelve years old, but the younger one could only just see over the hedge. Now, the elder boy was very much aware of the difference between the gold chain of the Mayor and the silver one of the Deputy.

"Poof!" said the older boy with contempt, "it's only the Deputy."

The little boy, with his eyes fixed on me in wonder, as I turned round, said, "Ooooo, I loves Deputies." It was obvious that he had seen many Westerns.

* * *

May 15th, 1962 was a very important date for me and, I might say, rather a nerve-racking occasion, for I was to be inducted by the Aldermen and Councillors as Mayor of Abingdon in front of invited guests. It was very difficult sorting out who should and should not be invited to the ceremony. Naturally my friends and relations had priority and they were allocated the front seats. Then followed a long list of important people like the Lord Lieutenant, the late Hon. David Smith, and his wife; the Member of Parliament, the late Airey Neave, and his wife; Group Captain and Mrs Neil Cameron from the RAF Station at Abingdon, and a great number of people drawn from all the societies, schools, associations, etc.

The ceremony took place in the beautiful Council Chamber, amid a wonderful floral display. My wife, Kitty, was very nervous; in fact, I had quite a job to persuade her to take part at all. Her first reaction to my being made Mayor was to threaten to go to Canada and stay with our son, Michael and that Jenny should be Mayoress instead of her.

I must say that I enjoyed thoroughly, my year of office, but there were times when, at full Council meetings with the Press and the general public present, political arguments would be voiced, and, as I was a member of the Independent Party, I had to tread very carefully.

In the one year, I attended over three hundred and sixty functions, dinners, balls, opening bazaars, receiving guests from overseas and entertaining in the Mayor's Parlour, etc. The great event of the year for the town, was the Mayor's Supper and Ball to which were invited the Mayors from the towns of the Thames Valley.

When I first joined the Abingdon Borough Council, I had heard that Abingdon had a fine display of Corporate Plate. Some said that, outside London, Norwich and Exeter, ours was the best and something to be really proud of. Having lived at Youlbury with Sir Arthur, this sort of thing interested me, and, on making enquiries, I was told that this silver was locked away and that I had to have permission from the Mayor to see it. After getting his permission, I then went to the Sergeant-at-Mace for the key and he took me into a small room which contained two very old safes which any burglar could have opened with a tin opener. What struck me most, when opening the safes, was that everything was all higgledy-piggledy, in fact, rather like a junk yard. I could see straight away, that we had a very fine collection of beautiful silver, including tankards, candelabra, and silver maces, dating back to Elizabeth I. Being very much a 'new boy', I kept my mouth shut, but, from time to time, some of the silver was cleaned and used on official occasions by the Mayor of the day. When I became Deputy Mayor, I began to make plans as to how the general public could see these treasures at any time without going through the Mayor. I had also seen, over the years, that every Mayor left some souvenir of his year of office.

I decided to put into action a scheme for housing the Corporate Plate in the Roysse Room. At the far end was the old broom cupboard, which was just what I wanted. The question then arose as to how the money was going to be raised, as I knew that the Council would not put up all the money. I did know, however, that, properly put to the general public, I would get the money I wanted, which was £1,200. I sent a letter out to the general public, banks, business houses, schools, the press, etc., explaining what I wanted to do with the Corporate Plate for, after all, this valuable treasure belonged to the burgesses of Abingdon. It is interesting to note that the first person who came to the Mayor's Parlour to see me concerning the appeal, was Group Captain Neil Cameron, (later to become Sir Neil and also Marshal of the RAF and head of all NATO air forces). He thought my idea was excellent and promised to send me a cheque, which was, I think, for £400. The following day, I arranged with Bert Edwards, the Sergeant at Arms, to remove all the Corporate Plate from the safes and arrange a display on built up shelves in the Roysse Room. The exhibition lasted for three days and to ensure its safety, Bert Edwards and his assistant, George Austin, slept on the floor, armed with pickaxe handles! But the only intruder was a large rat.

I made a special appeal that all the schoolchildren from the town should be brought down with their staff and I would give them a short resumé of the history of the Plate. One morning, some young children arrived and I stood them in front of the Plate while I stood to one side. As was usual on these occasions, I was wearing my gold chain of office. One small girl gave one glance at the display, which was surrounded by a bank of beautiful flowers, and then turned round and just simply stared at me with her large brown eyes. This became a little disconcerting and I began to think that I was not properly dressed. After my little discourse, I asked the child what she thought of the silver and the flowers, and all she could answer was, "Oh, I have never seen a real live Mayor before."

The Borough Council allocated me £300. Then, one day, an agent from ITV came to see me in the Parlour. He said he was prepared to give me a cheque for £200 if I would give him permission to run thirty new Mini Minor cars round the centre of the town, under a temporary arch in the High Street, constructed of flowers, with the words, 'Kellogg's Corn Flakes'. The cameras would be so placed that the County Hall would be in the background. The whole run of the Minis would need only five minutes. I thought £200 was well worth it, so I called in the Chief Superintendent of Police, John Snowley, who agreed on the five minute run and, furthermore, would hold up the traffic, on the condition that the run took place on a Thursday afternoon, when it was early closing. The general public were very amused when afterwards the ITV agent pulled out his cheque book, knelt down on the pavement and wrote me out the cheque.

Within two weeks, over £2,000 came in and work started straight away on building the showcase which was to be made with Triplex bandit-proof safety glass and connected to the local police station.

Meanwhile, I was most fortunate in securing the help of Mrs Pat Russell, who is very well known for her embroidery of church vestments. She undertook the arrangement of all the items, dating from Elizabeth I to the present day, inside the new display cabinet. On the panels outside the cabinet, were numbered all the items and their dates beside them, in beautiful calligraphy, in which Mrs Russell is also an expert. The unveiling of the Town's Corporate

Plate in the Roysse Room, on February 8th 1963, was by the Hon. David Smith, Lord Lieutenant of Berkshire.

Soon after joining the Council, I became aware that there seemed to be great arguments concerning the Abingdon mace and the one in use at the House of Commons, as to which one was the finer. Both maces were made by the mace maker, Thomas Maundy, and they certainly looked almost identical, but we, here in Abingdon, considered our mace to be a pure mace. One day, while in London, Kitty and I decided that we would go to the House of Commons and ask to see the mace. All sorts of obstacles were put in our way, even when I said that I was the Mayor of Abingdon. Not long after this, I happened to be attending a dinner at one of the colleges of Oxford, and, to my delight, I was seated next to Airey Neave. Of course, I brought up the question of the Abingdon mace and how I had tried to see the House of Commons mace but had been frustrated. Mr Neave also was aware of this long controversy about the two maces and said, "Well, Mr Mayor, I shall see what I can do." Of course, I thanked him, but I did not expect anything would come of it, his being a very busy man. You can imagine my surprise when I received a letter from him saying that the Speaker of the House of Commons would be pleased to see me, my wife, the Town Clerk, Mr E. Nicholson, and my Sergeant-at-Mace on February 11th, 1963. Off we set with the mace in a green bag, for the House of Commons. Mr Neave was there to welcome us and we then moved on to the terrace overlooking the Thames. There, of course, the Press were waiting and they asked to see our mace. Mr Edwards took it from its green bag and put it on his shoulder, for all to see. We then all went to the Speaker's private suite, where we met the Keeper of the Crown Jewels, General Sitwell, and the Deputy Sergeant-at-Arms, Lieut.-Col. Peter Thorne, who was an expert on maces. First, we had a glass of sherry, then we went into another room and there, on the table, was the House of Commons mace alongside the Abingdon mace. You could see, at once, that they were almost identical, but our mace which was made of silver-gilt weighing eight and a half pounds, was a little longer, four feet three inches. General Sitwell, on examining both maces, declared that the Abingdon one was a 'pure' mace, having been made by Thomas Maundy between May and September, 1660. The House of Commons one was also made by him, but there were some doubts about the head, as all maces throughout England were broken by order of Oliver Cromwell, and, after the Restoration of Charles II, Maundy had quickly to find a suitable mace for the House of Commons. Our mace, then, was the first whole one to be made after the Restoration. Some time later, Mr Neave told me that the Speaker had rapped him on the knuckles, for uncovering another mace within the precincts of the House.

It certainly was a very hectic time carrying out all the duties of a Mayor, even for one year.

On one particular occasion, I was asked by the Berkshire County Council, to open the new Old People's Home, in Ock Street, Abingdon, as a memorial to our first Mayor, Mayott, in 1556. On all official occasions, I had Mr Edwards to drive my Rover car. Kitty and I were suitably dressed and I was wearing the heavy, long gold chain. On arriving at Mayott's home, I was met by dignitaries from Reading, and the Warden of the home. We entered, amid the clapping of the inmates and Kitty was presented with a bouquet of flowers by the eldest inhabitant, aged ninety. After inspecting the buildings, we were given tea and I had to say a few words. Mr Edwards was waiting outside in the main street with the car, and after taking our leave in the proper manner, Mr Edwards opened the car door and we stepped in ready to set off, with smiles on our faces. To our complete surprise, the self-starter refused to work. I whispered to Mr Edwards to get out and swing the handle which he did, but with no result. Meanwhile, some of the crowd began to titter and I realized that I had to do something. I had had many years of swinging car handles during the war and I jumped out, took the car handle and gave it a real good swing. What I had not realized was that in swinging the handle, my long gold chain had wrapped itself round my neck and that I was about to strangle myself. I went to give the handle an even bigger swing, when Mr Edwards insisted that I must remove my gold chain. I was about to do this, when he stopped me, at once, by saying, "Mr Mayor, you cannot remove your chain here in the street, please come inside." By then, most of the people were in fits of laughter. Even inside, Mr Edwards insisted that removing the gold chain must be done in private. In the end, this protocol was carried out in the gents' loo! When I came out, again I had another go on the handle, but still no response from my car. The trouble was that the battery was flat and the only way to get the car going was to push it. Kitty sat in the back, Mr Edwards at the wheel and myself, with three other helpers, gave the car a good push. Mr Edwards slipped in the gear and the car started, accompanied by a great cheer from all the bystanders, as, by then, there was quite a crowd. They were all tickled to death to see their Mayor pushing his car.

In July 1962, I received an invitation from the Lord Chamberlain for Kitty, Jenny and myself to attend a Garden Party at Buckingham Palace. Off we set, with Mr Edwards driving the car, flying the small Borough flag that I used on certain ceremonial occasions. We drove into the forecourt of the Palace, an attendant led us through part of the Palace and then out into the gardens beyond. What a sight met our eyes! A huge crowd of people of every nation milling around in morning dress, uniforms or lounge suits, but what really caught the eye most, were the ladies' dresses which were of every colour under the sun. The Diplomatic Corps was very much in evidence and the ladies from overseas, dressed in their national costumes, made up a very interesting and striking part of the scene. Members of the Royal Family joined the guests while the Queen and Prince Philip took a stroll down the roped off path which twisted down the large lawn. From time to time, the

Queen and the Prince would stop for a few minutes to talk to various groups of people. No one was allowed to get under, or cross, the guide ropes. It was great fun picking out the diplomats, members of the Government and other famous people from all walks of life. After a while, we took ourselves off to one of the marquees for a tea of strawberries and cream. Then, for a change from being with the crowds, we wandered round the gardens and here we saw a beautiful sight, for, there in a large lake, was a group of flamingos and, just behind them, were a dozen Beefeaters (Yeoman of the Guard) dressed in their scarlet uniforms, marching along with their halberds on their shoulders; a sight never to be forgotten. After a time, a bell suddenly started to ring, which was a sign for everyone to leave. What a terrible noise it made, and when we came closer, we saw it was fixed to a corner of the Palace, in full view of everyone. It had a long rope and a member of the Palace staff was pulling it. It sounded just like a village school bell calling the children into school after they had been playing. When we got into the forecourt, we had to wait until our car was called – the Mayor of Abingdon! We all felt like poor Cinderella after the ball. Outside the Palace, were crowds of onlookers, with their faces pressed against the railings. You should have heard their comments on the women's dresses. What a change, coming from that world of Royalty, lovely gardens, strawberries and cream, straight into the London streets!

Life went on, day after day. Every morning, I had to be at the Mayor's Parlour to discuss with my secretary, my mail, what my appointments were and what functions I had to attend. Any spare time I had, I devoted to my own Company.

I remember one particular day which was very full. In the morning, some members of the town wanted to see me in the Parlour to ask me why they could not have a council house, why Mrs So-and-So had got a house when they, themselves had been on the waiting list for two years – always a dodgy one, that! Then, at 11 o'clock, I received a deputation from Oxford, of six African Chiefs from Swaziland. I was rather disappointed that they were dressed in lounge suits, as I would have liked it better had they been in their native dress. They seemed to like the sherry I gave them, but conversation was difficult. By the end of the day, I had had enough, but I still had to attend a sherry party at Roysse School, given by Mr Coban, the Headmaster. When I arrived home, I told Kitty that I would have a hot bath. As it was winter, it was dark outside and the curtains were drawn. I always have my bath before going to bed, and I don't know what happened, but when I got out of the bath that evening, I put on my pyjamas. At that point, Kitty came into the bedroom and greeted me by saying, "Good God, what are you doing? We have a sherry party at the school in half an hour." I wonder what would have happened if I had not had a wife!

The highlight of the year was the Mayor's Ball, which was always a glittering occasion and all the stops were pulled out. The flowers were gorgeous, there were two sittings for supper and dancing until after 1 a.m. Kitty looked absolutely wonderful in her evening dress and was a great credit to me.

I also had small private dinner parties in the Council Chamber or the Roysse Room, where the Corporate Plate was in full view. Only the family and close friends were present, Mr Edwards being on duty in the Parlour, serving pre-dinner drinks. It was at one of these parties in 1963, that we announced Philip's engagement to his first wife, Jilly. Kitty and I went to many balls in various towns and it was at the Royal Borough of Windsor that I had to respond to the speech by Princess Alice of Athlone, in 1963. Here, I must thank the Abingdon Rotary Club, for when I first joined them, I was scared stiff of saying even a few words, even to thank the weekly speaker. Now they cannot stop me!

Now, I must relate a 'clanger' that I made at the Mayor of Reading's ball. One day in 1962, when I was Mayor, the Lord Lieutenant, the Hon. David Smith, arrived at the Parlour as I was to accompany him on a visit to the RAF station. He arrived wearing full military uniform, red stripes down his trousers, a red band round his military hat, wearing a row of medals and carrying a swagger cane. As he was over six feet tall, he looked very impressive and I found him very easy to talk to. Off we went to the RAF station and met the CO Neil Cameron. The flying display was very impressive and, when we left, I asked the Lord Lieutenant to have tea in the Parlour with me. I saw him again, this time dressed in 'mufti', when he came to see the Corporate Plate in its new surroundings. Just before my term of office ended, I went with Kitty to the Mayor of Reading's ball. Kitty was, unfortunately, in another room and I was just leaving the Mayor's Parlour, when I bumped into a very tall man, dressed in white tie and tails, who at once knew who I was, as he asked me how I was getting on in Abingdon. I thought hard, but could not place him and then, after some conversation and noting his medals, I said, "Ah, yes. You are at the RAF station at Abingdon."

"No," he replied, "I am the Lord Lieutenant." I wished the ground could have swallowed me up. "Never mind, Mr Mayor. I am used to these sort of situations." He roared with laughter, so putting everything right.

At last, in May, 1964, I was able to step down for the new Mayor, Alderman Joe Stanley. I must say I was glad to finish my year of office; it had been so exhausting, and when you do finish, you find that there is nothing so dead as a past Mayor.

* * *

In August, 1963, our eldest son, Michael, was married, in Kingston, Ontario, to a very nice Scots girl, Maureen, who had emigrated from Scotland just after Michael. The following year, 1964, Philip was married to Jilly, in London, where he was working with the Express Dairy.

I stayed with the Borough Council for another two years in order to finish my period as an Alderman. I had, during this time, been a Governor of Christ's Hospital and Roysse School.

In 1966, Kitty and I spent a very happy holiday in Greece, visiting all the important archaeological sites right up to Meteora and then, of course, we went to Crete to visit the excavations at Knossos, and there, I had my photograph taken beside a fine statue, put up by the citizens of Crete, of Sir Arthur.

In 1967, we had two weddings; the first was our daughter, Jenny, to Peter Benson who is from a farming family I have known all my life. He is a corn merchant in Abingdon. The wedding took place in St. Helen's Church. The second was David to Mhorag Anderson. Mhorag, also a charming Scots girl, had lived for some time in Kenya and was an Air Hostess.

With all our family married, we could take even more holidays abroad, leaving the business safely in the hands of Mr Woodley. One of these holidays was spent in Norway with a party of Rotarians from this area. I was, at the time, President of the Abingdon Rotary Club. Soon after this, we visited all our relations in Canada, and David and Mhorag in Kenya.

In Kenya, we went on our first real safari. Mhorag, Kitty and I set off in the VW making for Tsavo Game Park on our way to Lake Manyara in Tanzania. Our first sighting of an elephant was in the bush, close to the road on which there was a sign saying, 'Elephants have right of passage'. We stopped to have a look, but did not linger long, as Mhorag said that as it was by itself, it might be a rogue elephant and, therefore, not to be trusted. We reached the Manyara Game Park, overlooking the lake, where we stayed for a couple of nights. Poor Mhorag could not eat her breakfast in the mornings because she was feeling sick.

After two days, we went on down to the lake, accompanied by an Askari to guide us through the many rough tracks with bush and trees on either side. We saw a number of elephants, sheltering under the trees from the heat of the sun. All at once, as we came round a bend, the road crossed an open space, and, to our complete surprise, we found ourselves in the midst of a large herd of elephants. Some of the herd had already crossed and the rest, including some babies, were about to follow. I do not know who was the most surprised, the elephants or us. Suddenly, it was all confusion and I was terrified. The bulls lifted their trunks, flapped their ears and trumpeted their anger at finding this little car, filled with humans, on the road. Meanwhile, the females gathered their young into a tight circle. The bulls seemed to be towering above us; it was very frightening and I expected to be flattened to the ground, but Mhorag never batted an eyelid. She realized that we could not reverse the car, so she put her foot down hard on the accelerator and we shot forward, clearing the herd. Kitty seemed to enjoy this encounter, but not I.

The following day, we went on to Ngorongoro to a very fine Game Lodge, 1,000 feet up above the crater, or caldera. Here we stayed the night and dined off zebra steaks, the zebra having been culled by the game wardens in order to keep the numbers down; and very nice they were, too.

Next morning, we had to use a Land Rover to descend into the crater by a narrow twisting road that had been built by Italian prisoners of war in the 1939-45 war. What a fantastic place it is; over nine miles across and thirty in circumference and leading on to the Serengeti Plains. There was such a variety of animals; anything up to twenty thousand head of game, which included hyenas, rhinoceros, wildebeest, lions, leopards and many others. Our great disappointment was not seeing any lions, which, we were told, were numerous in this area. All day we hunted for one, but not a sign did we see until it was time to return and climb the crater. Then, in some thick reeds, we had a glimpse of one lion, fast asleep on his back, like a kitten. Apparently, they normally sleep in the acacia trees during hot weather, to avoid the flies, but, owing to a recent storm, that had cooled the air, they were sleeping in the tall grass and the reeds.

On our return to Kenya, we stopped at Tanga, where Mhorag went to see a doctor. She came out of the surgery with a large smile on her face and told us what we had already guessed; she was going to have a baby. We then went home, picked up David and set off, in the small hours of the morning, for Entebbe in Uganda.

Within half an hour, I started to feel very sick. The roads were very rough indeed and, after a time, I was sick several times. I can remember looking up into the trees and seeing the monkeys watching me. David and Mhorag were rather worried and they decided that I should see the Mission doctor, a Scotswoman who lived in the valley at the Mission. We went to the hospital, where I was put in the waiting-room which I shared with several native women, a host of children and, to my amusement, a number of hens that pecked round me for food, and some small pigs that ran in and out, chased by the children. It seemed that I was suffering from car sickness and disorientation from travelling by night.

We went to Entebbe and then on to Murchison Falls, where we crossed the White Nile on a barge. It was dusk and our lights were reflected in the eyes of the many crocodiles that were watching us from the water, as we crossed. They did give me the creeps.

On arriving at the lodge where we were to stay, we found the guests dressed in evening clothes, standing at the bar where a white coated bartender was shaking cocktails. What a contrast it was, after spending all day in the heat among the animals and nature in the raw, taking just one step into that room. The accommodation, food and service were excellent.

Feeling tired, after our journey, we took ourselves off to our chalet, on the way encountering two hippos eating the grass on the lawns. The following morning, we had our breakfast on the terrace of the hotel, with only a low wall separating us from three elephants and a baby one that kept putting its trunk over the wall, begging for titbits. Now and again, the mother elephant would nudge the baby away from us.

All that morning, we travelled down the Nile in launches to watch the elephants and hippos and the many frightening looking crocodiles that were sunning themselves on the banks, with their huge jaws wide open and little birds working round their teeth for scraps of food. How it reminded me of my dentist. In order to get really close to the crocodiles on the bank, we shut off the engines and drifted past them. I could almost touch one very large one, but when he saw us so close to him, he slid into the water in a very sinister way. The hippos, asleep in the shallows, made a terrible splash diving into deeper water, as our boat approached. We would then just wait, making no noise, and, after a time, they would come up to breathe, thinking we had gone. Once or twice we got a good bump as they came up and hit the keel which was rather disconcerting, to say the least; but our native Ugandan guide was very pleased to watch our reaction.

Next day we were off again, first to Fort Portal and then down to our last Game Park, which was the Queen Elizabeth. This park was off the beaten track and had only a small Game Lodge and when we arrived in the evening, we found, to our consternation, that all the rooms had been taken. There was nothing else to do but push on another hundred miles to a town called Mburara, a rather dismal place with one hotel which we discovered was run by an Asian. It was certainly not up to the standards of the Game Lodges and how it smelt! It was nothing more than a row of rooms, one storey high, with a wooden veranda running the whole length of the building. We booked two rooms, Kitty and I having the only room with a bathroom. When I went to look at the room, I opened the door of the bathroom; one look and the smell was enough for me; I quickly closed the door. We were in room No 5 and David and Mhorag in No 15, the last room of the line of bedrooms.

We had a supper of sorts and then took ourselves off to bed. All the bedroom windows faced the veranda and we were told that an Askari would be on watch all night. Kitty and I undressed, locked the door and bolted the one window. However, the smell was so bad that to get some fresh air, I left the window slightly open. We had two single beds, Kitty's bed being nearest to the window which was on our left.

I do not know how long I had been asleep, when I opened my eyes, looked towards the window and there, silhouetted against it, I perceived the outline of a man's head, shoulder and arms, the rest of his body being obscured by Kitty's bed. My hand stretched out towards the table between the beds and, watching the figure at the window which never moved, I switched on the light and saw a man inside the room; he must have come in through the window. I let out a scream, switched out the light and then switched it on again. Kitty woke up and said, "Jim, you're dreaming. There's nothing to worry about." I let out a howl and then Kitty saw the man within three feet of her and she started screaming. The Ugandan was wearing only a short pair of pants. I jumped out of bed to grab him before he could go off with our cameras and passports and money and, as I did so, I caught my foot in a stool in the middle of the room and fell against the concrete wall beside the window, just as the man stepped over the window sill, grabbing a spare blanket as he went. Meanwhile, pandemonium broke out all around as the inmates of the other rooms started shouting, "Murder! Robbers!" When I fell against the wall, I had hit my forehead and blood just flowed down my face and my pyjamas, on to the floor and the bed. I grabbed a white towel and tried to stem the blood, Kitty called for help and David and Mhorag came running down the veranda in great alarm, for they had heard that there had been a murder in No 5! David, seeing the deep cut and the still flowing blood, dashed off to fetch the car to whisk me off to the local hospital. But first, we had to go to the police station and the sergeant in charge had to go through the usual procedures, name, address, father's name, mother's name, length of stay in the country, a description of what had happened, and so on. David, however, insisted that we go to the hospital to see a doctor, after which, we would return. The man agreed, and we set off in the car with the blood-stained towel wrapped around my head.

The man on night duty led us all into a room where an autopsy had just been done. There was a pile of blood-stained clothing on the floor and the instruments were still out on the table. The sight of all this was too much for me and I began to feel sick. The man had gone off to fetch the doctor who had gone to bed, as it was 2 a.m. I told David that I had to lie down or I would be sick. In an adjoining room was an operating table under a very bright light. I lay down on the table and felt a little better. When the doctor arrived, he immediately put five stitches into my head, and a very good job he made of it. We then had to go back to the police station to make a full report on what the doctor had done.

The Sergeant said, "Mr Candy, you are a very lucky man to be alive."

I said, "That be damned for a tale." I told him that I had not come thousands of miles just to have five stitches put into my head.

He replied, "Mr Candy, had you grabbed his leg as he stepped out of the window, he would have used his panga, and you would have been killed. Here, in this country, if a robber is caught, it is a hanging case."

We returned to the hotel and David and Mhorag stayed in our bedroom until daylight; there was no sleep for any of us. When we got up, the first thing I did, was to go to the office and complain bitterly. I wanted to know where the Askari, who was supposed to be protecting us, had been. It seems he had been asleep. I then asked for compensation, to which the Asian agreed, but he had to obtain permission from the directors, who lived in Nairobi, and I would have to wait in the hotel for at least two days. That was enough for me and I told him that I would leave at once and that I would not pay for the rooms or the meal. The manager agreed to this, and not waiting a moment longer, we left. As we went down the main street, we saw a café which looked at least clean. We pulled up and went inside for breakfast. To my surprise, the owner was English and I told him what had happened. He told us that that hotel was known in the district as 'Dysentery Hotel'. Perhaps we were lucky, after all.

We went straight back to Nairobi and stayed with Mhorag's aunt and uncle, Jenny and Jim Smith of the Alliance High School. They looked after me so well and their doctor ordered me to stay in bed for twenty-four hours to get over the shock. We had meant to travel on to Addis Ababa to see the Coptic churches, but all I wanted to do was to return home for a quiet life.

We also travelled all over Europe again and, on one occasion, spent a very good holiday with my old Las Petacas friend, Frank Corkery.

We had another lovely holiday when we flew to Venice and then went by boat to Split, Yugoslavia, on the S.S. *Uganda*, down to Alexandria, from where we took a trip to see the Pyramids. Just before entering the port of Alexandria, we were told by the Captain, that, on no account, were we to take any photographs of the Egyptian navy, either from the decks or the portholes. As we passed along, we saw one or two ships and three submarines which seemed to be covered in rust and 'decked out' with the sailors' 'smalls' drying. I wonder what price a foreign power would have paid me for a photograph of this?

We visited the spectacular island of Santorini, where we disembarked and travelled by donkey one thousand feet up a twisting path to reach the top. It was this island that had a large part of it blown out of the sea, in 1750 BC. It was the biggest natural explosion recorded on earth and it destroyed the Minoan civilisation, which was not to be rediscovered until Sir Arthur Evans excavated Knossos in Crete, in 1900. We went to Crete again and stayed in Iraklion, visiting the museum and other Minoan sites. It was in Iraklion, in the 1980s that the Cretans demonstrated in large numbers in the streets, against the decision of the Greek Government, to send the famous Minoan treasures from Knossos and other sites, to Europe and America for exhibition. The treasures were theirs and they were not going to allow them to be shipped abroad. The Cretans won; the idea was dropped by the Government.

* * *

Now that all our children were married, all we had to do was wait for the grandchildren. Sure enough, along they came, much to our delight, and Kitty's knitting needles were soon working overtime. First on the scene was Paul, followed by Caroline, then Alison, followed a month later, by Lisa, then Natasha, closely followed by Fiona, next came James, then Mark, then Louise, closely followed by Julie, and finally, at the time of writing, Sally. What good-looking children they are – all eleven of them.

We went through a very unhappy time when Philip and Jilly's marriage broke up and they got divorced. Philip came to live with us and Kitty was a great help to poor Philip; she spent many hours talking and listening to him, sometimes not coming to bed until the small hours of the morning. In time, of course, he got over his troubles and in 1975, he married Anita, a delightful girl and a graduate of York University, and they have two children, Louise and Sally. Jilly has brought up Caroline and Alison very well indeed and it makes us very happy and how well they fit in with the rest of the family and their cousins.

My sister, Madge, died from cancer in 1973, while Kitty and I were in Tenerife. In 1958, Madge, one day after having her bath, felt a very small lump in her brest and within an hour, she was with her doctor. He advised her to have an operation and she went into hospital that very day. Sure enough, it was cancer and her breast was removed. For the next fifteen years she enjoyed a perfectly healthy life and was never ill. It just shows, if cancer is taken in time, life can be prolonged.

In 1973, my brother, Dick, who had been badly wounded in the First World War, died from a heart attack. He had worked for Canadian Forestry for thirty years. I had always felt very close to Dick, especially during our Blagrove days. It is a great delight for me to be with Claude, either here or in Canada, to talk over old times. I think I was half in love with her myself when she and Dick were engaged.

In 1973, Kitty and I had a delightful holiday in Bermuda with Michael and Maureen and the children. Michael had been sent there by his bank. What a delightful island, but I can quite see that, after a time, one could have a feeling of claustrophobia.

In 1974, we went to Canada with a party of Rotarians, organized by John Proctor. On arriving at Vancouver, we were given a very good welcome by the Rotary Club which had laid on two coaches to take us, for two weeks, round the Rockies. What a spectacular journey that was and how the Rotary Clubs, in the towns where we stayed, entertained us. Kitty and I had been through the Rockies before, once by train and then flying over them on our way to Vancouver.

In two weeks, we travelled two thousand miles amongst the most fantastic scenery. Every night, we stopped at certain towns and were hosted by the local Rotarians. Their hospitality and kindness was out of this world, the only drawback for me was that they kept us up long after my usual bedtime. You could not blame them; not only did they wine and dine us, but they were avid for news of England. Concerts and balls were laid on everywhere and there was one especially good ball at Bamff. It was in Bamff where we met the wild brown bears that wandered around the hotel, dining off the scraps from our dinner plates.

Somehow or other, I caught a cold that settled on my chest, so I took the opportunity, as we were staying two whole days with our hosts, to see a doctor. I was given some medicine and some rather large pills to take twice a day to ward off any other infection that I might pick up.

We made our way towards Vancouver through the Kicking Horse Pass with its five hairpin bends, which I did not like as there were no guard rails and a 1,000 feet sheer drop. At one particular hairpin bend, the driver decided to take the 'mickey' out of us. This bend was impossible to take in one sweep and we had to back towards the edge with its sheer drop to the river. He told us that he had never taken passengers up the Pass before, only goods, and he asked the passengers at the rear of the coach to tell him how far he could go back, saying that we could all be strawberry jam, if we miscalculated. Then he started messing about with the gears, making out that he could not get into bottom gear. This was really too much for me so I left my seat and told the driver that I wanted to leave the coach. He realized that he had overdone it. He told me to return to my seat and he locked the doors. Many of us had had the 'wind up', but I received a ticking off from Kitty for trying to leave the coach without taking her with me. We discovered, afterwards, that he was a very experienced driver.

Our last official stop of the tour was for a luncheon at a small town quite close to Vancouver, but, by then, I was not feeling very well and was longing to get to our hotel and rest. At last, late in the afternoon, we arrived in Vancouver and were taken to the famous Rose Gardens.

Whilst everyone was admiring the gardens, I suddenly felt very ill indeed so I asked Kitty to tell the driver to stop and telephone for a taxi. I just could not carry on any longer in the coach. We got out and sat on a garden seat while the driver telephoned. I sat there, just holding my head. I did not feel sick or have any pain anywhere and I heard one or two of the Rotarians whispering, "It must be his heart"!

When we arrived at the hotel, I just flopped on to the bed and, after an hour, I felt a little better. Thank goodness it was the end of the trip; the next day our Rotarians were flying back to England while Kitty and I still had another two weeks visiting relations. Kitty had telephoned my nephew, Peter Candy, who lived in West Vancouver, and he said that he would come over in his car and take us home where his wife, Sheila, was cooking a nice turkey dinner for us. When Peter arrived, I told him that I just wanted to stay where I was till morning, but he thought I ought to see a doctor so he took me to the hospital which was quite close by.

At Reception, I told the receptionist that I was feeling ill and would like to see a doctor. She sent me into the room next door where, behind a large typewriter, sat a real old 'battleaxe' of a woman. She wanted to know my name and address, my age, was I married, how many children I had, if my parents were alive and, if not, what had they died of, and so on. I was feeling ill again, so I said, "Look, all I want is to see a doctor." She said that she had to have all these vital statistics before I could see one. How this reminded me of the police station in Uganda. These formalities over, I was told to sit down and wait for the sister and Peter said he would wait in the car.

After a little while, the sister arrived and asked me to follow her and she led me into a large ward containing many beds, a good few of which had green curtains round them. The sister went to a cupboard and, to my horror, produced a large white nightshirt and a pair of long white stockings. You can imagine how I felt. They must be going to operate on me. What had I got? Oh dear, what would Kitty say? I was then led to a bed and told to undress. "Fold up your clothes and put them on the chair with your shoes on top," she said. This I did and, after a time, she came back and pulled up the sides of the bed from underneath so that it was impossible for me to get out, and then she took my clothes away.

Now, from my bed, I was able to look round this large ward, and among the patients were the usual road accident victims and so on, but I soon became aware that from the curtained beds, came the sounds of slapping and the voices of nurses saying "Come along, now, Mr So-and-So, wake up," followed by the sounds of more slapping. Poor devils, I thought, they must have been operated on and the nurses were trying to bring them round.

After a time, a very pretty nurse came to see me. She took my temperature and enquired about my bowels. She asked me where I came from and, as I knew it was no use saying Abingdon, I said Oxford, which is only five miles from Abingdon. We chatted and I began to feel better. Then off she went and brought back another nurse, even prettier. She also, asked me where I came from and, when again I replied Oxford, the first nurse said, "There, I told you so. You can tell by his accent."

Meanwhile, I could still hear the sound of slapping and, "Come along, Mr So-and-So," from at least six different beds. Poor things, I was surprised that the surgeons had been so busy on a Saturday.

After about half an hour, a white-coated doctor arrived. He let down the sides of the bed and sat down and proceeded to tap me all over, listen to my heart and test my blood pressure. I was becoming increasingly apprehensive, wondering if they were going to operate on me. After a while, he pulled up the sides of the bed and said, "I am sending for another doctor to have a look at you." Good God, I thought, this is serious. There must be something very wrong. Soon, the other doctor appeared and, to my surprise, I saw he was Chinese and under five feet tall. He let down the sides of my bed and sat down, saying that he had been sent by the other doctor. Having spent over an hour in bed and been cheered up by the two pretty nurses, I was feeling much better now.

He asked me several questions and then asked me if I had been on a trip. I replied, "Oh yes, doctor, I've been on a marvellous trip. One of the best. Fantastic sights, all round the Rockies; a world of its own; never to be forgotten."

He asked me to sit up and he took out a torch from the pocket of his white coat. "Will you close one eye and keep the other open? Now let me see into the other eye." When he had finished, he said, "Mr Candy, there is nothing wrong with you; all you need is a day in bed as you are suffering from exhaustion. I will discharge you and tell the sister to bring your clothes."

As soon as my clothes arrived, I was out of that bed, dressed and away to find poor Peter. Even as I was dressing, the same curious smacking noises could be heard accompanied by the voices of the nurses, pleading for their patients to wake up. Then, suddenly, like a bolt from the blue, the realization came to me that they were drug addicts, and they had thought I was one, too. I remembered how I had enthused over my 'trip' to the doctor. Peter was most relieved to see me and when I explained what had happened, he roared with laughter and said, "Of course, this always happens here on a Saturday night when they are picked up from the streets."

We returned to the hotel where poor Kitty had been waiting and wondering and I went straight to bed while Kitty and Peter went off to have supper with Sheila. How they laughed over my story and, in fact, I became known as 'Jimmy the Junkie'. The following morning, after a good night's sleep, we set off, by boat, for Victoria to stay with my sister-in-law, Claude and to see my sister, Ethel.

In 1975, we went for a few days to Seville, where, for the first time since leaving the Argentine in 1933, we aired our Spanish. Every time we spoke Spanish in the hotels and shops, they spotted, at once, that we came from South America.

Through the Rotary Club, I heard that John Proctor and his wife, were organizing another tour, but, this time, it was to New Zealand. This was just what I wanted as it would give me a chance to see my brother, Gilbert, and his wife, Esther, whom I had last seen in England, in 1950, and to meet the numerous members of his family.

The tour was an extensive one and on our outward journey, we stayed for two nights in Singapore, a very interesting and clean place, quite different from all the other countries in Western Europe. We then set off for New Zealand and a tour of Rotary Clubs in South Island. What hospitality we received. At Queenstown, they took us, by bus, to the River Shotover where they dished us out with shovels and pans and we spent the whole afternoon panning for gold. One or two of our members did find a speck of gold the size of a pinhead. South Island reminded me of Scotland and when we crossed by boat to the North Island, what a contrast it was. I wished Kitty had come with me, but she had decided that, as she had never met my New Zealand relatives, she would much prefer to visit David and Mhorag in Kenya. She was always shy when meeting people she did not know.

I must say, I did like North Island with its beautiful scenery. Everything was so green and there was so much to see; its thermal activity, its pools of boiling mud, just like porridge when it starts to boil, and most spectacular of all, the displays of the geysers erupting from the rocks in great quantities of boiling water, every few minutes, high up into the air. The tour lasted three weeks and during that time, I ran into six people from Abingdon.

In 1977, we took a holiday in Austria, one of the countries we had not visited before. It was in early spring, a delightful time of the year, with the snow still lingering in the shady parts and the marigolds in full flower in the little streams.

Not long after we returned, Kitty discovered a small lump in her left breast, which she had removed at the Ackland Hospital in Oxford, and I was told that she had cancer.

As soon as she had recovered from the operation, we went to stay with Michael and Maureen in Toronto. This was a wonderful holiday as Michael took us, by car, down to Washington to see Aldo Raffa, our old friend from our early travels in Greece. We then went on to Williamsburg, Jamestown, Charlotsville and finally to Abingdon, Virginia.

Here, the Mayor of Abingdon gave us a wonderful reception. He owned the Mary Washington Hotel and, to our amusement, he allocated the Bridal Suite, complete with a large bowl of fruit, to us. He would address me as, "Mr Mayor," but I told him that I had 'died' fifteen years ago. He replied, "Once a Mayor, always a Mayor."

That night, we had seats given to us at the Barter Theatre to see *The Mousetrap*. The manager of the theatre came on to the stage before the show, to enquire who had come the furthest distance from Abingdon, Virginia. It turned out that we had and, much to the delight of the audience, we received a mug which had been made in the town. I asked the Mayor why the theatre was called the Barter Theatre and he told me that during the depression in the thirties, people could not afford the price of a seat, so the Theatre Company decided to keep going by the simple method of bartering the seats for chicken, eggs, vegetables, fruit and, in fact, everything that could be eaten. In this way, everyone was happy. Even to this day, on the opening night of the season, they still carry out this tradition. I would love to be there for an opening night.

We really enjoyed that visit with Michael and Maureen, especially the ride through the Blue Mountains.

When we arrived back in England in early October, Kitty seemed to be very well, but just before Christmas, I noticed a change in her and she often felt very tired. On Boxing Day, she was not at all well and had pains in her arms and legs. After a visit to the Specialist and having a bone scan, it was discovered that she had cancer in her bones.

During her illness, if the weather was good, we would go out for drives in the car. How she loved to go round the Cotswolds and have lunch in the village pubs and watch the fish and ducks at Bibury.

In the spring of 1978, Kitty became worse and had to stay in bed all the time, but she had one delight when Jenny brought little Julie to see her every day, after dropping Fiona and Mark off at school. While Jenny went off shopping, Julie would sit on Kitty's bed and play with her beads, without a word of protest and never once did she attempt to get off the bed. How Kitty looked forward to her visits each morning.

At the beginning of May, Kitty was taken to the Marcham Road Hospital where she died on the 13th. She was cremated on the 18th, after a service at Christchurch, Northcourt.

From the day she knew that she had cancer, she never complained, nor did she ever talk to me about her illness – what courage! She was a wonderful wife. I must remember Dr Peter Webb for the way he looked after her, never missing a day without a visit while she was ill at home.

* * *

I still continued to work for the Company, but in a very small way. In 1978, we had amalgamated with Cliffords Dairies at Bracknell. This had been an important step for we were now dealing in thousands of gallons per day. Philip was responsible for the whole retail section of the company and there were many further purchases of small dairies and, in some cases, large gallonages. I remember, many years ago, telling Kitty that I had met Sam Farrant on my rounds and that he wanted to sell me his little milk business which consisted of five gallons a day, as he was a large dairy producer and did not want to be involved in milk distribution. I agreed to buy, as it was in my area, and I paid him £5 for it!

Then followed a strange event which brought me back to Sir Arthur. A certain Miss Dorothy Matthews, who lived on Boars Hill, came to see me to ask if I could be persuaded to join the Boars Hill Oxford Preservation Society. The local committee wanted to erect a suitable plaque to the memory of Sir Arthur, in the Wild Garden surrounding Jarn Mound, which Sir Arthur had planned and had laid out. Miss Matthews knew of my many years connection with him. I agreed and at the very first meeting, Dr Bredin, who was Chairman of the committee, asked me if I would undertake finding a suitable memorial. In a flash, I seemed to know exactly what would have pleased Sir Arthur – a very large natural stone. One of the first things he had done when I was a little boy and living with him, was to take me to the Rollright Stones, Avebury and Stonehenge, all places he loved.

The next day, I went down to the offices of Curtis and Sons in Abingdon, to ask if they had any large stones from their gravel pits and, sure enough, they had one in their yard from their pit at Tackly. On seeing the stone, I knew, at once, that it would have indeed pleased Sir Arthur, for it was irregular in shape and weighing over ten tons. I explained why I wanted it and Mr Curtis Junior said that I could have it, free of charge, and that he would transport it to Jarn Mound and also send a crane which could lift the stone right over the cherry trees to where it was to be erected. He told me it would cost nothing as Sir Arthur had given him plenty of work in the past and that it was a P.R. job. I went straight back to Dr Bredin and Miss Matthews who were delighted with the whole scheme.

Then came the problem of what wording should be written on the plaque that was to be inserted in the stone. Now, this did cause a problem but it was left for me to deal with and I was to bring what I thought would be appropriate before the committee.

Before the stone was actually erected, Dr Bredin resigned and Mrs Marian Stevens became Chairman and she soon formed a committee of very enthusiastic members of Boars Hill residents.

In time, I did draw up some wording for the plaque but it did seem to me to be much too long and I thought that people would not stop to read it. The committee agreed but asked me to persevere and then I had the bright idea of going to see Dame Joan Evans, Sir Arthur's half-sister, whom I had known in the old days at Youlbury, so I telephoned her to see if she could help me. She invited me down to her house, at Wotton-under-Edge, for lunch.

When I arrived, I was taken into the drawing-room, where she sat, nursing an old cat on her lap. She was nearly blind and unable to walk without help.

"Well, Jim, let me see what you have written about Arthur," she said and, having read it, she went on, "do you see that torn piece of paper? Please pass it to me." Then, giving me her old cat to nurse and, without the slightest hesitation, she dashed off the following: *Arthur John Evans, 1851-1941, who loved antiquity, nature, freedom and youth, created this viewpoint and wild gardens for all to enjoy.* What a gift for words! I asked her if she would unveil the plaque and she said she would if transport could be provided and if she was well enough after a cataract operation on her eyes.

There was some delay in deciding who was to do the carving on the plaque which was made of Cumberland slate but, eventually, it was done by Michael Harvy who did an excellent job. Soon after it was finished, Joan Evans telephoned me to say that she was too ill to unveil the plaque. I then consulted our committee and we decided to ask the Greek Ambassador, His Excellency, Mr Stavros Georgiou Roussos to do it and when asked, he readily agreed. Marian Stevens asked me to contact Joan Evans again to suggest a suitable person to introduce the Greek Ambassador. She at once suggested Professor Hood, who had excavated at Knossos, but not in Sir Arthur's time, or Professor Warren, as he, too, had recently worked at Knossos and other parts of Crete. In fact, Joan Evans became very ill and died a few months before the unveiling ceremony which took place at Jarn Mound on June the 17th, 1978.

Many people attended, not only local residents, but others from the Ashmoleum Museum, Oxford, from London and the local press. Marian very kindly invited many of them to tea in her lovely garden on Boars Hill. Just after tea, Marian asked me to say a few words to the guests concerning Sir Arthur, particularly his human side as most people were aware of what he had accomplished during his long life, especially his discovery of the Minoan civilization.

I spoke for about ten minutes giving my audience some idea of the preoccupations that governed this extraordinary, kind man.

Little did I know that there was, amongst the guests, a woman, unknown to me, who listened with great interest to what I said and about three weeks later, I received a letter from Jerusalem, from the American authoress, Sylvia Horwitz, asking me if she could come and see me to discuss Sir Arthur as she was writing a biography of him, (she had worked at Knossos) and she had heard of me through this woman who had been at the unveiling and the tea-party. I wrote back at once saying that I would be delighted to meet her and, within a few days, Sylvia arrived.

I took her up to Jarn Mound so that she could see what the Boars Hill Committee had achieved and after lunch, we sat down on the terrace and she tape recorded all I had to tell her about Sir Arthur and my life with him. She was fascinated by everything I had to tell her and she incorporated some of it in her book and she was delighted with some photographs that Dennis Haskins sent her. It is strange to think that Dennis and I are the only two people alive in the world today that really knew Sir Arthur. All that day and far into the night, we talked about him and looked at the one hundred and thirty letters he had written to me from 1913 until 1941. Her book is called *Find of a Lifetime* and is published by Weidenfeld and Nicolson. She has also written biographies of the painter, Goya and Toulouse Lautrec. Sylvia was most emphatic that I should write my memoirs, as I had had an interesting and varied life, and in her letters to me during the last two or three years, has encouraged me in every way to keep writing and I have sent her copies of the draft to New York, for her comments.

Another strange twist came about three months after the unveiling of the plaque. I had a visit from my old friend, Aldo Raffa, whose wife had died of cancer just two months before Kitty. In the course of conversation, I told him about Sir Arthur and Jarn Mound and I then went on to say how a certain Sylvia Howitz had encouraged me to write about my life. When I mentioned her name, he jumped up from his chair exclaiming, "My wife and I were very great friends of Sylvia and her husband over thirty years ago while I was working with UNRRA during the war. I thought that they must both be dead. Howitz was in charge in Milan and I was restoring order in Florence." Aldo went to my desk, took a piece of writing paper and dashed off a note to Sylvia. You can imagine how surprised and delighted she was when she received the letter for she, too, had thought that the Raffas must be dead.

Encouraged by Sylvia's insistence that I should start writing, I decided to have a try, but I knew that as I was not very scholarly, I was going to be 'up a gum tree'. Help came from a most unexpected quarter.

I must take you back to 1933, when, soon after our return from the Argentine, Kitty and I went to visit my old friends, Sir Frank and Lady Madge, at Dormans Park, East Grinstead. What a lovely weekend for us both. Such a happy home: a pleasure to be with them all. Frank and Woggie took such delight in showing us the twins June and Doreen. They took us up to the nursery where, to our amusement, we found them sitting on their potties, before coming down to lunch and that night, Frank insisted on showing me how to bath them, in preparation for the future! Our association with the family has never been broken, for although their parents have long been dead, we have never lost contact with the girls and Doreen, or Dozey, as she is nicknamed, had continued to come and see us from time to time after she was married and living in Bewdley, Worcestershire. After a time, her marriage broke up and she moved to Prestbury in Gloucestershire, with her three young adopted children. Although I was fond of all the girls, I knew Dozey the best and her intelligence, sensitivity, love of the arts and appreciation of the countryside, made her a wonderful companion. After Kitty died, David and Mhorag invited us both out to Kenya to stay with them. And so it was to Dozey that I appealed for help. This was forthcoming at once, so in January 1980, I started to write. Dozey did the correcting and then Jenny did the typing, but, after a time, Jenny had to give up the typing as, besides having three children and livestock to look after, she helped Peter in his office. So, apart from some typing that Dell Mellor very kindly did for me, Dozey has done the lot, and a good job she has made of it, too. Little did I think as I bathed her all those years ago, that she would be helping me today.

Early in 1982, I contemplated visiting my relations in Argentina, as I had not been back since leaving with Kitty in 1933, but, after some consideration I decided that going to see my brother, Gilbert, in New Zealand, and my sister, Ethel and her husband, Fred, in Canada, had top priority. Gilbert was approaching his ninety-fourth birthday and Ethel was not far behind. How right I had been in not going to the Argentine as I would have landed just two weeks before the invasion of the Falkland Islands.

On March the 24th, I set off by plane, 'round the world in eighty days'. I did not realize until I returned home, that it was exactly eighty days from start to finish.

I spent the first few days with my old friend, Frank Corkery in Portugal, and then flew on to Australia and New Zealand, visiting my relations. It was nice to see Gilbert and Esther and all the rest of the Candys. I was very thankful that I saw my brother, for, a month after I returned to England, he died.

I then set off for Vancouver, visiting Fiji on the way, and stayed with Sheila and Peter. It was lovely to see everyone again, especially Ethel and Fred and their daughter, Sally; they were all looking so well. Ethel, at ninety years old, was still playing bridge three times a week.

I also took a short trip down the Oregon coast with Connie Heston, an American woman living in Spokane, near Seattle. I had known her and her family since 1975 when she stayed in Abingdon, having been given a year's course at Oxford by the Rotary International Fellowship to study retarded children.

I then returned to Vancouver, where I had asked Dozey to join me, and we had a marvellous time staying with my relations there and with Claude and also my niece, Margaret, in Victoria on Vancouver Island. We then flew to Toronto and stayed with Michael and Maureen, after spending a very pleasant few days with my nephew, Allen and his wife, Norma. This was a delightful holiday and we both very much enjoyed ourselves.

Dozey flew back to London and I went on to Washington to see Aldo Raffa and to New York to visit Sylvia. Here, again, I had more encouragement to finish my story and to let Sylvia have a copy of what I had so far written. After a trip to Charlottesville, I returned home on the Concorde.

On September 27th, 1982, my family gave me a marvellous eightieth birthday party. Unfortunately, Michael and his family were not there and James was at school too far away to come, but the other eight grandchildren were all there. We had drinks in the main office of the company and a very good cold lunch in Philip's office. Amongst the guests were Mary and Jack Cloney, Winnie Crow, Dell Mellor, Gwen Moss, Elsie Mullard, Elsie and Arthur White, Peter's parents, Phil and Frank Benson, Anita's parents, Peggy and Eric Hoare, Dozey Thursfield and Marian and Dallas Stevens. All these people had been great friends of Kitty's and mine for many years, our oldest friends being Marian and Dallas, who were married in 1933, after us and had come to live on Lady Singer's estate, within a hundred yards of our first home in England.

After lunch, we all came up to Monks Way to cut the cake and drink a toast in Champagne. What a happy day!

On November 30th, 1982, I resigned from the company, after sixty years of work.

It is strange, but as I get older, I tend to sit down by the window and let my eyes range over the garden that holds many sweet memories and my thoughts start to travel back to Blagrove days when, on a Saturday, Dick and I would watch the old path that led to Oxford, anticipating the arrival of Frank Gillams. When I saw him coming, I would run to fetch the spade and Jock, the dog, would go mad with excitement as he knew we would be off rabbiting. Travelling not so far back, I can still hear Sir Arthur's voice on the telephone saying, "Er, er, er, is that you, Jimmie? Would you care to sup with me tonight?"